3D Modeling and Printing with
TINKERCAD®
Create and Print Your Own 3D Models

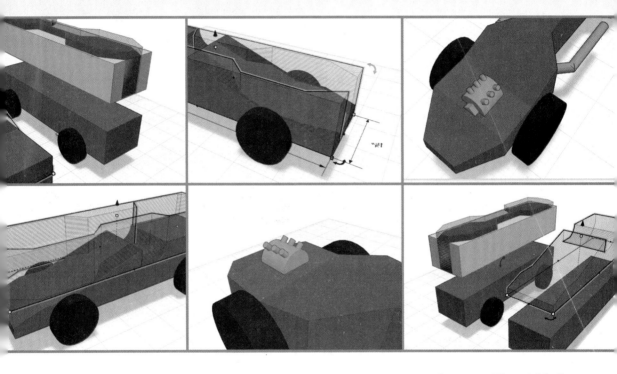

James Floyd Kelly

THE BRYANT LIBRARY
2 PAPER MILL ROAD
ROSLYN, NY 11576-2193

que®

800 East 96th Street,
Indianapolis, Indiana 46240 USA

3D Modeling and Printing with Tinkercad®

Copyright © 2014 by Pearson Education

All rights reserved. No part of this book shall be reproduced, stored in a retrieval system, or transmitted by any means, electronic, mechanical, photocopying, recording, or otherwise, without written permission from the publisher. No patent liability is assumed with respect to the use of the information contained herein. Although every precaution has been taken in the preparation of this book, the publisher and author assume no responsibility for errors or omissions. Nor is any liability assumed for damages resulting from the use of the information contained herein.

ISBN-13: 978-0-7897-5490-5
ISBN-10: 0-7897-5490-8

Library of Congress Control Number: 2014934912

Printed in the United States of America

First Printing: June 2014

Trademarks

All terms mentioned in this book that are known to be trademarks or service marks have been appropriately capitalized. Que Publishing cannot attest to the accuracy of this information. Use of a term in this book should not be regarded as affecting the validity of any trademark or service mark.

Autodesk, the Autodesk logo, AutoCAD, Tinkercad, and Autodesk 123D are registered trademarks or trademarks of Autodesk, Inc., and/or its subsidiaries and/or affiliates in the USA and/or other countries.

Warning and Disclaimer

Every effort has been made to make this book as complete and as accurate as possible, but no warranty or fitness is implied. The information provided is on an "as is" basis. The author and the publisher shall have neither liability nor responsibility to any person or entity with respect to any loss or damages arising from the information contained in this book or from the use of the programs accompanying it.

Special Sales

For information about buying this title in bulk quantities, or for special sales opportunities (which may include electronic versions; custom cover designs; and content particular to your business, training goals, marketing focus, or branding interests), please contact our corporate sales department at corpsales@pearsoned.com or (800) 382-3419.

For government sales inquiries, please contact governmentsales@pearsoned.com.

For questions about sales outside the U.S., please contact international@pearsoned.com.

Editor-in-Chief
Greg Wiegand

Executive Editor
Rick Kughen

Development Editor
William Abner

Managing Editor
Kristy Hart

Senior Project Editor
Betsy Gratner

Copy Editor
Kitty Wilson

Indexer
Erika Millen

Proofreader
The Wordsmithery LLC

Technical Editor
Ralph Grabowski

Editorial Assistant
Cindy Teeters

Cover Designer
Mark Shirar

Compositor
Nonie Ratcliff

Table of Contents

Introduction 1

Chapter 1   **3D Modeling Is Cool!** 5
   What Is 3D Modeling? 5
   Where Have You Seen 3D Modeling? 11
   Why Is 3D Modeling Useful? 13
   What Can You Do with 3D Modeling? 14

Chapter 2   **3D Modeling Basics** 17
   What Are Axes? 17
   What Is a Plane? 22
   Understanding Rotation 25

Chapter 3   **Say Hello to Tinkercad** 29
   Finding and Opening Tinkercad 30
   Navigating Tinkercad 35
   Changing a 3D Model's Properties 40
   Looking at Lessons 43

Chapter 4   **Learn Some Modeling Basics** 45
   The Launchpad 45
   The Rocket's Main Body 53
   The Rocket's Fins 57

Chapter 5   **Putting Together a Model** 69
   Assembling the Launchpad 70
   Assembling the Rocket 86

Chapter 6   **A Tinkercad Special Project** 101
   Brainstorming Ideas 102
   Creating the Basic Tag Shape 104
   Adding Embellishments 119
   Adding Raised Text 122
   Suggestions for Improvements 126

Table of Contents

Introduction 1

Chapter 1   **3D Modeling Is Cool!**   5
     What Is 3D Modeling? 5
     Where Have You Seen 3D Modeling? 11
     Why Is 3D Modeling Useful? 13
     What Can You Do with 3D Modeling? 14

Chapter 2   **3D Modeling Basics**   17
     What Are Axes? 17
     What Is a Plane? 22
     Understanding Rotation 25

Chapter 3   **Say Hello to Tinkercad**   29
     Finding and Opening Tinkercad 30
     Navigating Tinkercad 35
     Changing a 3D Model's Properties 40
     Looking at Lessons 43

Chapter 4   **Learn Some Modeling Basics**   45
     The Launchpad 45
     The Rocket's Main Body 53
     The Rocket's Fins 57

Chapter 5   **Putting Together a Model**   69
     Assembling the Launchpad 70
     Assembling the Rocket 86

Chapter 6   **A Tinkercad Special Project**   101
     Brainstorming Ideas 102
     Creating the Basic Tag Shape 104
     Adding Embellishments 119
     Adding Raised Text 122
     Suggestions for Improvements 126

Chapter 7	**Another Tinkercad Special Project 129**	
	Developing an Idea 130	
	Creating a Mold for the Object 138	
	Creating the Elements for the Mold 143	
	Finishing Up the Mold-Making Project 147	
Chapter 8	**Printing Your 3D Models 155**	
	What Is a 3D Printer? 156	
	Creating an STL File 161	
	Melting That Plastic 165	
	Moving the Nozzle 167	
	Using Software to Control a 3D Printer 171	
	Summary of 3D Printing 175	
Chapter 9	**More Useful Tricks with Tinkercad 177**	
	Using the Mirror Feature 177	
	Importing Your Own Sketch 187	
	Experimenting with the Shape Generators Tool 194	
	Where to Next? 197	
Chapter 10	**Where Can You Find Existing Models? 199**	
	Welcome to Thingiverse 199	
	Additional 3D Model Sources 209	
Chapter 11	**Expanding Tinkercad's Usefulness 211**	
	Finding a 3D Printing Service 212	
	Taking Your Object into *Minecraft* 217	
Chapter 12	**Special App for Turning Real-World Objects into 3D Models 225**	
	Converting Real Objects to Digital Models 225	
	Improving Your 3D Modeling Skills 241	
Appendix A	**More Free CAD Applications to Explore 243**	
	123D Design 243	
	SketchUp 244	
	FreeCAD 245	

Appendix B A Bonus Project 247

The Pinewood Derby 247

Creating Digital Body Shapes 250

"Carving" the Block 259

Appendix C A Closer Look at 123D Design 267

123D Design Interface 267

The Main Toolbar 269

The Navigation Bar 276

What's Left? 283

Index 285

About the Author

James Floyd Kelly is a writer from Atlanta, Georgia. He has degrees in industrial engineering and English and has written technology books on a number of subjects, including CNC machines, 3D printing, open software, LEGO robotics, and electronics.

Dedication

For Decker and Sawyer: "Who wants to do a project?"

Acknowledgments

I love writing books that help others, especially younger readers. I'm a jack-of-all-trades, master of none, and I frequently have to teach myself new subjects that interest me. This means hunting down material that's often vague, hidden, or incomplete…sometimes all three. Because of this, when I choose to pass on my knowledge in a book, I do my best to organize it in a way that makes sense to me and, hopefully, my readers.

Along for this trip are a number of key individuals who at various points in the writing process provide me with help, feedback, and support…sometimes all three.

First, my sincere gratitude goes to Rick Kughen, who saw the potential in my early proposal. He offered advice on what to include and what to cut that really did make the book much better.

Next is my technical editor, Ralph Grabowski. A technical editor is supposed to help catch my errors and tell me when I've forgotten a step or maybe need to be more specific when describing something. Ralph did a great job of finding my mistakes and suggesting improvements. Any additional errors you find belong to me. (And please let me know if you find any by emailing feedback@quepublishing.com.)

I also must thank five more folks at Pearson: Laura Norman, development editors William Abner and Todd Brakke, managing editor Kristy Hart, and my project editor, Betsy Gratner, for helping to keep me organized. If you enjoy the book, all these individuals deserve a large part of the credit.

Finally, I must thank my wife for her support and patience and my two young boys, who keep me always on the lookout for new and fun projects to tackle.

James Floyd Kelly

Atlanta, March 2014

We Want to Hear from You!

As the reader of this book, *you* are our most important critic and commentator. We value your opinion and want to know what we're doing right, what we could do better, what areas you'd like to see us publish in, and any other words of wisdom you're willing to pass our way.

We welcome your comments. You can email or write to let us know what you did or didn't like about this book—as well as what we can do to make our books better.

Please note that we cannot help you with technical problems related to the topic of this book.

When you write, please be sure to include this book's title and author as well as your name and email address. We will carefully review your comments and share them with the author and editors who worked on the book.

Email: feedback@quepublishing.com

Mail: Que Publishing
        ATTN: Reader Feedback
        800 East 96th Street
        Indianapolis, IN 46240 USA

Reader Services

Visit our website and register this book at quepublishing.com/register for convenient access to any updates, downloads, or errata that might be available for this book.

Introduction

Welcome to Tinkercad!

About 20 years ago, I went off to college to study engineering. One of the classes that engineering students had to take was a drawing course that required us to use special pencils (not the familiar No. 2s) and rulers and straight edges to create hand-drawn schematics of various objects. We started out with simple designs like cubes and pyramids and moved slowly but surely into more advanced drawings. It was a fun class, but it was also very tedious. I could sometimes erase mistakes, but many times it was easier to just start over, especially when I discovered I'd made a measurement error early in a complex drawing.

At the end of the class, the instructor informed us that a special class was being offered the next year, called CAD, or Computer-Aided Design (although some students called it Computer-Aided Drafting). He said that to get in the class, a student would have to demonstrate some basic skills. As an example, the professor taped three large drawings on the wall; each of the drawings showed one particular face of a large cube—a top, a side, and a front view. This cube consisted of 9 smaller cubes on each face, just like a Rubik's Cube, as shown in Figure I.1.

These drawings, however, showed that the cube had a few missing smaller cubes, which created a strange landscape on a few of the faces of the larger cube. Students had 60 seconds to draw a version of this 3D cube using nothing but the front, side, and top views. The professor was testing our ability to visualize a 3D object in our head, based on three simple views, and then transfer that image to paper.

I didn't take the test because I wasn't interested in the class. But the students who did take the CAD class said that it was both fun and difficult (a typical description of many engineering classes) and that it involved drawing objects on a computer screen by defining points on the object and then connecting lines between points. It sounded difficult, time-consuming, and frustrating...but most of the students said it was much faster than drawing by hand, as we'd all done in that beginner-level class.

Today's CAD applications are much more advanced than those that existed 20 years ago. These applications are colorful, they do much more of the work for you (such as drawing a perfect circle or ensuring that a line is exactly 1.275 inches in length), and they allow you to create 3D objects that exist on a screen long before they exist as real physical objects.

INTRODUCTION

FIGURE I.1 A Rubik's Cube.

Many people complain that learning to use CAD applications is difficult. I've tried to teach myself half a dozen CAD applications over the past few years, and although I've had success with them, they've been frustrating at times, and mastering them requires a serious time commitment. Until very recently, anyone who wanted to use these applications to create 3D objects had to put in the time, deal with a lot of frustration, and often pay a substantial fee to get the software if their school or workplace didn't provide it.

But not anymore. Today, a few CAD applications are tailored to beginners and less intimidating than their big-brother counterparts. They require less time to master and in some cases are even free to use. One of these CAD applications is called Tinkercad.

I first encountered Tinkercad at Maker Faire (www.makerfaire.com) back in 2012. I was hooked by a number of factors: It was impressive and colorful; using it required only an Internet connection and a web browser; using it felt almost like dragging and dropping LEGO blocks onto the screen; and it was free. I created an account, began to use Tinkercad, and discovered that it's a pretty good CAD app. It has limitations, but for a beginner-level CAD application, it's an impressive example of how a complex tool can be simplified for anyone to use.

What you're holding in your hands is a book for learning Tinkercad. The popular CAD software company, Autodesk (www.autodesk.com), makers of the commercial and very popular CAD application Autodesk 360, purchased Tinkercad in 2013, and the company chose to keep the application free to use—good news to the fans who had already discovered the simple CAD app.

Autodesk didn't just buy Tinkercad; the company has continued to improve the app, adding new tools and features, and providing users with free online support via a blog and forum.

If you're new to CAD applications, this book is for you. If you understand the importance of a CAD application but aren't sure where to start, this book is also for you. Maybe you're the brand-new owner of a 3D printer and are wanting to start designing and printing objects to print in plastic—if so, Tinkercad is the perfect tool to get started. (And check out Chapter 8, "Printing Your 3D Models," if you're not familiar with 3D printers and want to know more.) And if you've found Tinkercad but are feeling a little confused about where to start, you've got the right book. I'm going to show you all you need to know to begin creating some amazing 3D objects, and hopefully when you're done with the book, you'll feel confident enough to continue on with any new features introduced to Tinkercad or maybe even move on to playing with a more advanced CAD application.

Using Tinkercad truly is one of the easiest ways to experience computer-aided design and to create 3D models. You're about to learn a new skill...and you're going to have a lot of fun doing it. I'll see you in Chapter 1, "3D Modeling Is Cool!"

3D Modeling Is Cool!

In This Chapter
- What is 3D modeling?
- Where have you seen 3D modeling?
- Why is 3D modeling useful?
- What can you do with 3D modeling?

Are you familiar with the term *3D modeling*? If so, feel free to jump to the next chapter, but I'm hoping you'll stick around and continue reading because this chapter talks about so many different examples of 3D modeling that you might discover something new and interesting.

If you're not familiar with the term, you will be by the end of this chapter. You're holding a book that is going to teach you how to use a fun (yet powerful) piece of software called Tinkercad. Using Tinkercad is easy and fun, but before you begin working with it, it will be helpful to understand some of the things you can do with the software (as well as some of the things you cannot).

This is a fun book, with lots of hands-on projects for you. But before you can run, you've got to learn to walk, right? That's what this chapter is all about—getting you standing and ready to walk. Let's go...

What Is 3D Modeling?

Take a moment to examine the toy car shown in Figure 1.1. (Even better, if you've got a toy car of your own, go ahead and grab it.)

Take a look at this car and consider some of its characteristics. The car has four wheels. It has a front and a back and two sides. It's longer than it is wide. Its height is close to its width, but we'd probably need a ruler to get exact measurements. It has a certain weight to it, and it's probably made of plastic or metal or both.

CHAPTER 1: 3D Modeling Is Cool!

FIGURE 1.1 A small toy car held in my hand.

One thing is for certain: The car is a physical item that can be picked up and examined. I can look underneath it, I can spin the wheels, and I can roll it across the floor. This car is real; it's a tangible item I can hold in my hands.

Now take a look at the same car in Figure 1.2 and compare it to the toy car in Figure 1.1. What details can you determine about this car?

The car shown in Figure 1.2 is flat because it's being displayed on my tablet's screen. I can't pick up this car and examine it as I can the real car. I don't know how much this car weighs or what it's made of. It's really nothing more than an image on the screen, isn't it? I could take a ruler and measure its height and width, but even that information can't be trusted. Look at Figure 1.3, and you'll notice that I've zoomed in on the car: The height and width have changed slightly.

There are a lot of differences between the real car and the car on the screen, but one key difference between the two is that one takes up real, physical space and the other takes up nothing more than a few megabytes of digital storage space on my tablet.

Another way to look at this is that one car is real, and the other is a representation of a car. It looks like a toy car, but it's really nothing more than a replica...a simple model.

What Is 3D Modeling? 7

FIGURE 1.2 A toy car on my tablet's screen.

FIGURE 1.3 I've zoomed in on the image of the toy car.

CHAPTER 1: 3D Modeling Is Cool!

There are different definitions of the word *model*, but the one I'm most interested in here is "an example or imitation of a real object." Maybe you've seen model airplanes or model cars that you snap or glue together. These models are miniature representations of larger original objects. If a teacher asks a student to make a model of the Eiffel Tower, that student (hopefully) isn't going to try to build an exact, full-size replica of the Eiffel Tower on the playground. Instead, the student might decide to make a smaller model out of toothpicks or cardboard. In this example, a real-world object is re-created on a smaller scale; it can be transported from house to classroom in Mom or Dad's car.

Another interesting thing about models is that they can be made first, before the real thing. Engineers at Ford or Toyota often create new car designs out of clay first, at a much smaller scale.

These clay cars are typically half the size of the final cars, or maybe even smaller. It's much cheaper for a car designer to experiment and test the shape and look of a car by first creating a clay model. If the model is approved, the larger, real car is finalized and put into production for people to buy and drive.

This process of taking an existing object (or the idea of a real object) and first creating a model of it is called—you guessed it—*modeling*. Modeling, therefore, is simply creating a representation that can be used to stand in for a real object, whether it be a car, a toy, a building, or something else.

But modeling is only half of the term *3D modeling*. Fortunately, 3D is an easy concept for us humans. 3D, as you know, stands for *three dimensional*, but there are also a few other dimensions you should know about. A point in space is said to be *one dimensional*, or 1D. A point lacks length, width, and height; it's just a point. Anything that has length and width is said to be *two dimensional*, or 2D. A good example is your desktop. It's flat and has length and width. Place a 2D piece of paper on that desktop and draw a square. That square is a 2D object. It has no height.

NOTE

A drawn 2D object technically does have height, but it's very difficult to measure. The pencil lead or ink from a pen that is deposited on a piece of paper when you draw a square is sitting on top of the paper, and that square could be argued to be a 3D object because it has height and width and an almost imperceptible height from the lead or ink. For all practical purposes, however, a drawn object is said to be a 2D object.

As you've probably figured out, a 3D object has length, width, and height. Look around, and you're going to see dozens of objects that exist in three dimensions—chairs, cups, bags, books, and so much more. You live in a three-dimensional world!

Interestingly, a 2D object can sometimes represent a 3D object. Look at Figure 1.4, which shows a drawing of a cube sitting on a flat surface. The cube looks 3D because of the way it has been drawn. This is sometimes called an *orthogonal view*, which just means you're looking at an object from a certain angle that allows you to see two or more sides of the object.

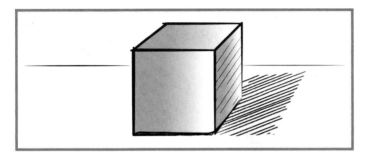

FIGURE 1.4 This cube is still a 2D object.

Now here's the interesting part: As you can see with the cube in Figure 1.4, a 3D object doesn't necessarily have to be a real object. You can hold a real object in your hands and examine it. Think about the real toy car. You can rotate it in your hands and look underneath it, you can roll it on the floor, and you can even drop it in a glass of water to see if it floats. (Probably not.)

Now look again at that 3D cube drawing in Figure 1.4. Can you rotate it to see the bottom? No. Can you roll it across the floor as you would a pair of dice? No. Can you drop it in a glass of water? No. You could drop the piece of paper the cube is drawn on in the glass of water, but you can't drop the actual cube itself.

The drawing of the cube is static: You cannot hold it in your hands, and you most definitely cannot rotate it around to see what's on its nonvisible sides. But what if you could rotate it? What if you could turn that drawing into a real object that could be held in your hands?

Up to this point, the cube in question has existed only as a drawing on a piece of paper. But what if it could be drawn in such a way that you could rotate the actual drawing?

Fortunately, this ability exists, and it's possible using software that allows you (or anyone else who has the necessary skills) to draw objects on a screen and rotate them. The software allows you to shrink or enlarge an object, to change its look, to take what is in your head as an idea and re-create it on the screen as a 3D model.

If you used this software to re-create the Eiffel Tower, instead of having a toothpick model, you would have a representation of the Eiffel Tower small enough to fit on your computer or tablet's screen that you could rotate and zoom in or out; you could even change its look to suit your whim. A smaller model might not have the same level of detail as the larger original, but as you can see in Figure 1.5, it's still recognizable.

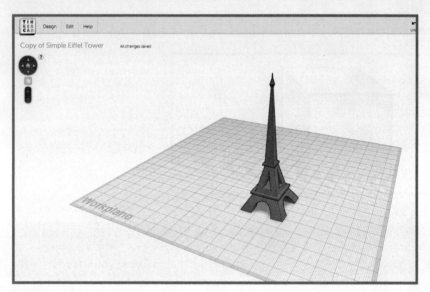

FIGURE 1.5 A 3D model of the Eiffel Tower. (Model created by Shawn M. and hosted by Tinkercad.com.)

With the right software, you can turn a 3D model into a physical model you can touch and hold, or a model that exists on a screen for viewing. Either way, you get a 3D model. And the process of creating that 3D model is called, yep, *3D modeling*.

NOTE

There are hundreds of different 3D modeling software applications out there, and you'll learn about a few of them later in this book. 3D modeling software is often also called computer-aided design (CAD) software, which basically means the computer helps create digital objects. (Okay, it's actually the computer *running* the software that helps.) Think about trying to draw a perfect circle or a straight line with a pen or pencil. CAD software allows you to select a tool to draw circles, for example, and then makes certain the circle on screen is perfectly round while leaving it up to you to determine its diameter (or radius). This will all start to make more sense once you start using the Tinkercad software. (Notice that *cad* is part of the name!)

Where Have You Seen 3D Modeling?

I've already mentioned the clay car model and the toothpick Eiffel Tower as examples of models that are created to represent larger objects. Models exist all around you. There are physical 3D models, and there are digital 3D models.

Have you seen an animated movie lately? The creators of all the *Toy Story* movies modeled hundreds of objects using software and then modified those models so they appear to move. Bringing 3D models to life in a movie requires high-power hardware and software, but most objects you see in an animated movie these days started out as nonmoving 3D objects created on a computer screen. Figure 1.6 shows a character being created using very specific (and complicated) 3D modeling software called Blender. The finished character will have height, width, and length and can be rotated in such a way that every point on its body can be viewed. (Don't worry. The software you'll be learning, Tinkercad, isn't this scary looking, and it's not difficult to use at all.)

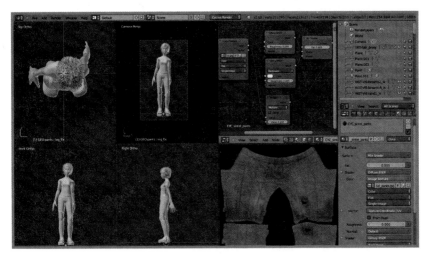

FIGURE 1.6 An animated character starts as a 3D model. (Image courtesy of Blender.org.)

Ever heard of *Minecraft* (www.minecraft.net)? It's an amazing and popular game that allows players to move around a digital island and build things like castles, cars, and more. Objects are assembled using the game's most basic building block: a cube. Cubes have different colors and patterns on them to represent what they are made of: rock, steel, and other materials. Figure 1.7 shows a scene from *Minecraft*, in which the player has built a house from cubes.

CHAPTER 1: 3D Modeling Is Cool!

FIGURE 1.7 *Minecraft* lets players create 3D models to use in the game. (Image courtesy of Minecraftmuseum.net.)

You may have played a video game that allows you to wander randomly around and view objects and characters and buildings from various angles. These are typically games that were designed using some sort of 3D modeling software. One of my favorite games is *Team Fortress 2* (www.teamfortress.com), and the characters that you see in Figure 1.8 were all initially created using 3D modeling software.

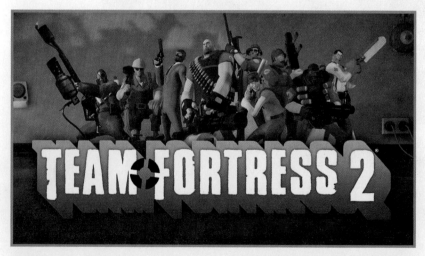

FIGURE 1.8 *Team Fortress 2* characters started as 3D models.

3D modeling is everywhere. Architects use it to design digital models of buildings that are then approved and built in the real world. Technology companies use 3D modeling software to create the look and feel of new products, such as the case that will protect a mobile phone. Artists use 3D modeling software to design jewelry and sculptures. Automobiles, airplanes, helicopters, and ships are all designed with 3D modeling software.

I hope you're beginning to understand just how many jobs and careers rely on a knowledge of 3D modeling. It's a skill that is useful to a wide variety of people, and the need for 3D modeling skills is only going to increase. Read on to find out why.

Why Is 3D Modeling Useful?

In 1890, building the Eiffel Tower cost 8 million francs, which translated to about US$1.5 million at the time. Today, that same building would cost almost US$40 million. It took more than 7,000 tons of iron to build and required 60 tons of paint. Its design existed on paper plans that construction workers referenced as they built.

That's a lot of materials and a lot of money. Can you imagine what would have happened if a mistake had been found halfway through the construction that had required a larger base or different angles for supporting beams? Fortunately, the designers were good at their jobs, as were the construction workers. Today, however, no government or company would ever agree to allow a structure so large (or even one half as large) to be built without first seeing a number of models built. Small models, medium-sized models, and even large models would be created, most likely from different materials. This would allow decision makers to look at the form, the color, and the structure for safety and strength needs.

Making physical models and making digital models both cost time and money, but once a digital model is made, it can be easily modified. It can be copied, and each copy can be given a different color or texture. A model can be stretched taller or squished into a shorter version. Imagine the time needed to create multiple similar physical models to display on a table!

There are many benefits to creating and using digital 3D models:

- **They are easily modifiable**—With software, you can quickly change sizes, colors, textures, and other characteristics of digital 3D models, usually with just a click and a drag of the mouse.
- **They don't take up physical space**—The screen that displays digital models may take space, but the model itself doesn't. You can zoom in and out on a digital 3D model and even display varieties of the same model on a single screen.
- **They are portable**—Imagine having to ship a toothpick Eiffel Tower from one side of the country to the other. Not only would it be expensive, but you'd have to pack it carefully to keep it from getting damaged. On the other hand, you can send digital 3D models electronically over the Internet or as email attachments.

- **They are testable**—Believe it or not, software exists that tests 3D models in simulated environments. For example, you can use such software to test 3D models of buildings to see how they would withstand hurricanes or earthquakes, you can test car models for wind resistance, and you can test plastic toys to see what they can endure and how well springs and other moving parts will hold up.

Digital 3D models can save time and money in the development of new products, in the design of buildings and vehicles, and in the testing of all these models in simulated real-world conditions. But there's one other thing you should know about creating 3D models: It's really fun.

What Can You Do with 3D Modeling?

While creating 3D models is really fun, I don't want you to forget just how valuable a skill 3D modeling can be. The ability to use software to create original objects or to re-create real-world objects in digital form is in demand, and you might find opportunities to use 3D modeling skills with a paying job. Yes, you can earn real money if you're good at creating digital models. Imagine that!

So 3D modeling might be a great skill for a future job. But what can you do right now with 3D modeling software? There are hundreds of 3D modeling applications available; some are free, and others cost $1,000 or more. Some are called 3D modeling software, some are called CAD applications, and others have names that don't even hint at their function. With the proper 3D modeling application, you can do some amazing things:

- **Design your own games**—Games like *Minecraft* allow you to create 3D objects and interact with them. Advanced skills with 3D modeling software allow you to create your own characters and worlds. Combine this with some computer programming skills, and you may be on your way to creating the next *World of Warcraft*!

- **Make your own movies**—3D modeling is the key to all the latest animated movies, and even live-action movies often use computer graphics (CG) for special effects. Imagine making your own home movie and inserting an animated Tyrannosaurus rex that chases you around the house.

- **Create a digital you**—With the right software, you can model yourself. And you'll learn later in the book about special hardware that can save you time by converting digital photos of you (or any other object) into 3D models that you can then import into 3D modeling software.

- **Print a real-world object**—As you'll find out later in the book, a 3D printer can turn a 3D model file that you create into a real plastic object that you can actually hold in your hand.

What Can You Do with 3D Modeling?

Before you can do any of these tasks, however, you need to not only find a good 3D modeling application to use, but you need to understand some basic terminology related to creating 3D models.

In Chapter 2, "3D Modeling Basics," you're going to learn some of the basic concepts that most 3D modeling applications expect you to understand. Don't worry. There's no advanced math, and the concepts are not very complicated. And when you understand these concepts, you'll be ready to take your first look at Tinkercad, 3D modeling software that is perfect for beginners.

2

3D Modeling Basics

In This Chapter
- What are axes?
- What is a plane?
- Understanding rotation

This is the chapter in a book that I always dread writing: the one about basic terminology. I'm going to keep this chapter short and sweet. Instead of throwing hundreds of terms and concepts at you, I'm only going to cover three. Yep, that's right: just three. Before you dive in and start learning Tinkercad, I'm going to explain these three simple ideas: axes, planes, and rotation. If there's anything else you need to know, I'll give it to you when you need it.

You may actually already have a solid grasp on axes, planes, and rotation. If you do, feel free to skip any or all of the sections in this chapter, but I hope you'll at least glance at those sections to make certain we agree on the terms.

What Are Axes?

Some people see the word *axes* and think I'm speaking of a collection of sharp, tree-cutting tools, like the one shown in Figure 2.1.

No, I'm talking about the plural of the word *axis*. The *ax* part is pronounced like "Max" or "Tax," and the *is* part sounds like "hiss" or "miss" and not a "zzzzz" sound. Say it out loud with me: "AX-IS"!

The plural of axis isn't axises or axi, but it's still a strange variation: axes. You already know how to pronounce the *ax* part, but the *es* part now sounds like "peas" or "freeze"—that's a long *e* there. Say it with me: "AX-EEEES"!

Axis and axes...the first means one axis, not two, not three, and most certainly not four. One. One axis. Anything more than one, and you've moved to axes. Two axes. Three axes. Four axes. Got it?

FIGURE 2.1 You won't be learning about this tool here.

I know you're nodding, but you're probably also asking, "What is an axis?" (Or maybe you're asking "What are axes?" Isn't English fun?)

Back in Chapter 1, "3D Modeling Is Cool!" you read about 1D, 2D, and 3D. The *D* stands for *dimension* or *dimensional* and is directly related to the concept of axes. Look at Figure 2.2, and you'll see a number line that runs right and left. You've probably encountered such a line in school. The numbers to the right of 0 are positive, and the numbers to the left of 0 are negative.

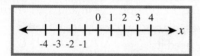

FIGURE 2.2 A horizontal number line that is called the X axis.

Now, look at Figure 2.3, and you'll see that a drawing has been placed above this number line. The axis in this case allows you to determine the width of the object.

FIGURE 2.3 A 2D object placed above the number line.

We refer to this number line that runs left and right as the *X axis*. The X axis can be used to specify the width of any 2D (two-dimensional, or flat) object. The object in Figure 2.3 has a width of three. But three what? Three inches? Three meters? Right now the units of measurement haven't been specified, but just keep in mind that real objects will always be measured using some form of units.

What Are Axes?

Now take a look at Figure 2.4. In this figure, I've added another number line that is running vertically on the page and is referred to as the *Y axis*.

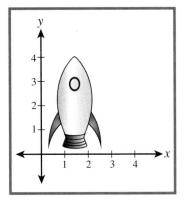

FIGURE 2.4 A new vertical number line called the Y axis.

The Y axis can be used to specify the length of a 2D object, so the 2D object in Figure 2.4 has a length of four.

NOTE

This measurement system is traditionally referred to as the *Cartesian coordinate system*. It's used to specify the location of individual points in 2D and 3D objects. Another system is called polar coordinates. For purposes of learning and using CAD applications (specifically Tinkercad), you'll be using the Cartesian system.

So, a 2D object can have its width and length specified using two number lines, called the X and Y axes. The X axis measures the width, and the Y axis measures the length. But what about a 3D object? It has another dimension: height.

Figure 2.5 shows a 3D object drawn on a flat surface that contains the X and Y axis number lines. You can see the object's width and length, and your eyes are probably telling you that this is indeed a 3D object, but how can you get the value of the object's height if you're looking straight down on the X and Y axes?

To find the height value, you need to add one more number line, called the *Z axis*. Where the X and Y axis number lines cross is called the *origin*. It is the 2D location where X=0 and Y=0. The Z axis pierces the origin like an arrow, going straight down into and through the other side of the flat surface. To see this new number line, you simply need to rotate the X and Y

CHAPTER 2: 3D Modeling Basics

axes so you're looking at them from a different angle. Figure 2.6 shows one possible viewing angle.

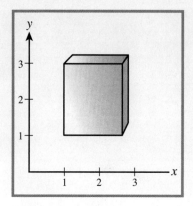

FIGURE 2.5 A 3D object placed on the X and Y axes.

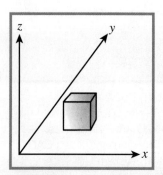

FIGURE 2.6 With the Z axis, a 3D object has a measurable height.

In Figure 2.6, you're looking at the 3D object from the side and slightly above. This vantage point allows all three axes to be visible and for your eyes to verify that the object has height, width, and length.

NOTE

If you've played any video games that allow you to rotate around an object (such as your player character or another game object), then you're probably already familiar with being able to view an object from multiple angles. When using CAD software, you use your mouse (or a laptop's touchpad or your fingers on a touchscreen) to change the viewing angle for an object on the screen.

The concepts of axis and axes will begin to make more sense once you're actually using 3D modeling software such as Tinkercad. Instead of using a numbered line, however, you'll be creating and modifying 3D objects on the flat workplane shown in Figure 2.7. This workplane can be rotated and viewed from different angles, and it's almost always used to represent the X and Y axes—but without the visible number lines.

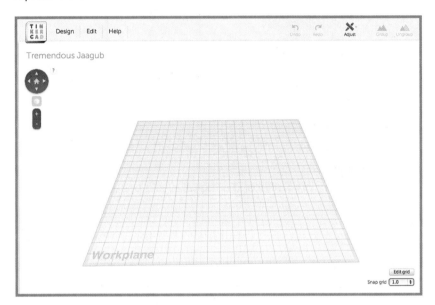

FIGURE 2.7 A flat digital workplane in Tinkercad.

You may have noticed that there is no origin designated on the Tinkercad workplane. There is no point on the workplane where X=0 and Y=0. The wide-open space on the screen is like a physical desktop on which you would write a letter or play a board game. And just as you'd use a physical (flat) desktop to create 3D objects out of clay or cardboard or other material, you use this digital workspace on the screen to create digital 3D objects. The digital workspace shown in Figure 2.7 can easily hold a 2D object that has only width and length, as shown in Figure 2.8.

By the way, this flat surface that consists of the X and Y axes goes by the name "workplane," and that leads us to our next 3D modeling concept: the plane.

CHAPTER 2: 3D Modeling Basics

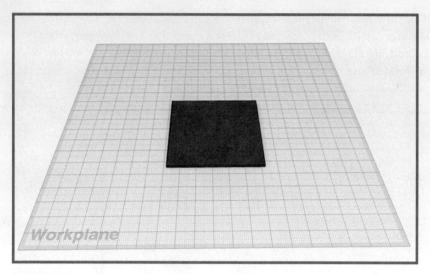

FIGURE 2.8 A 2D object resting on Tinkercad's digital workspace.

What Is a Plane?

Since I'm writing about 3D modeling concepts, when I'm talking about planes, I'm not talking about those objects in the sky that carry passengers from one place to another. Instead, I'm talking about an infinite flat surface that shares two axes.

Place a sheet of paper on the desk in front of you and draw a 2D object on it—a square, a stick figure, or maybe your design for a time machine. Whatever you choose to draw, that flat drawn object now exists on the X and Y axes. Another way to say this is that it sits on the XY plane. A *plane* is a flat surface that extends outward along two axes. Figure 2.9 shows an orthographic rendering of the XY plane sketched on a piece of paper.

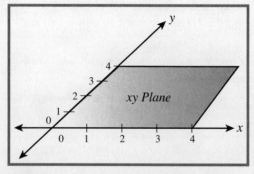

FIGURE 2.9 The XY plane is a 2D flat surface.

Notice in Figure 2.9 that this plane has an origin where X=0 and Y=0 in the lower-left corner. Now take a look at Figure 2.10, and you'll see that I've expanded the four edges of the XY plane so that it goes in every direction for the X and Y number lines.

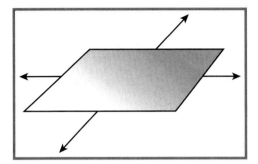

FIGURE 2.10 The expanded XY plane.

An infinite number of XY planes exist. How can this be possible? Imagine for a moment that the piece of paper sitting on your desk is hovering slightly above the desktop. To see this properly, you need to see the Z axis, as in Figure 2.11.

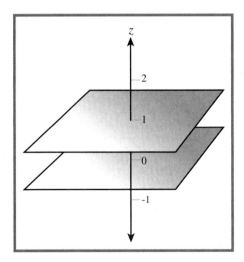

FIGURE 2.11 An XY plane where Z=1.

Back in Figure 2.10, the Z axis isn't visible; the XY plane shown there is sitting in the spot where Z=0. The flat surface that expands out on the X and Y number lines where Z=1 is another plane—but one that's sitting one unit above the plane where Z=0. XY planes can go up and up and up (where Z is positive)…or they can go down and down and down (where Z is negative). In case this isn't making sense yet, we'll go with another example to see if things begin to clear up.

Figure 2.12 shows the YZ plane. A 2D object placed on the YZ plane can be measured along the Y axis or the Z axis and exists perpendicular to the X axis.

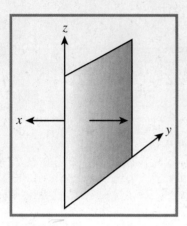

FIGURE 2.12 The YZ plane slices into the X axis.

And yes, there's an XZ plane as well, as shown in Figure 2.13.

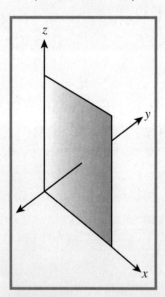

FIGURE 2.13 The XZ plane slices into the Y axis.

Planes become important when you design digital 3D objects onscreen. Take the cube shown in Figure 2.14, for example. This cube has six sides, and each side resides on a different plane. Two are XY planes (the top and bottom), two are YZ planes (the left and right sides), and two are XZ planes (the front and back).

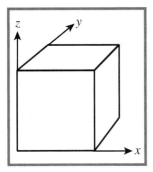

FIGURE 2.14 A cube's surfaces exist on six different planes.

As you begin to learn to use 3D modeling software like Tinkercad, you'll discover that sometimes you need to change the view so that a specific plane is displayed on the screen. This will become more apparent once you start creating 3D objects and rotating them around to view them from all angles (as well as make changes).

Understand that changing the view of a complete object on the screen isn't the same as rotating an object (or a piece of an object) with respect to another object or the workspace. Often while creating 3D objects, you will need to rotate a piece of an object while keeping all other pieces stationary. This is the next topic.

Understanding Rotation

For a moment, I want you to think of turning a key in a lock. Notice that the lock stays fixed, but the key rotates. With 3D modeling software, you'll easily be able to select an object (like the key) and rotate it in space while keeping everything else fixed in place (like the lock). This is one type of rotation that's fairly easy to understand, but there's another kind of rotation you'll be doing in Tinkercad (and other 3D modeling software) that involves rotating something on a specific axis.

Remember that a 3D object has three axes that help define the object's size and position. The X axis typically corresponds to an object's width, the Y axis to the object's length, and the Z axis to the object's height.

Look at the object in Figure 2.15 that's sitting flat on a desk.

There are many ways to rotate this object. You can turn it horizontally so that its flat bottom doesn't leave the table top. You can rotate it clockwise or counterclockwise while leaving it in exactly the same position on the flat surface, as shown in Figure 2.16.

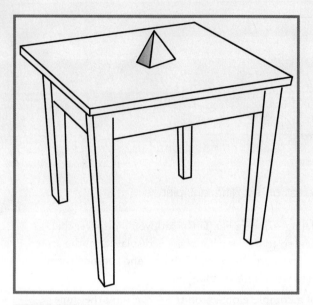

FIGURE 2.15 A simple object sitting on a flat surface.

FIGURE 2.16 Rotating the object left to right while keeping it in place.

You can also rotate it front to back, as shown in Figure 2.17.

All three of these rotations are done along a specific axis. Rotating the object in place is the same as spinning it along a Z axis. Imagine a Z axis number line entering the top of the object and exiting the bottom, as shown in Figure 2.18.

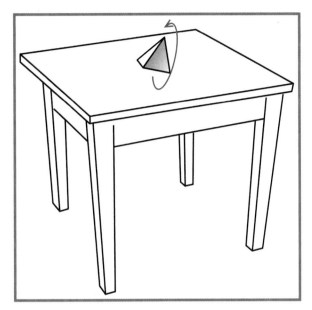

FIGURE 2.17 Rotate the object front to back while keeping it in place.

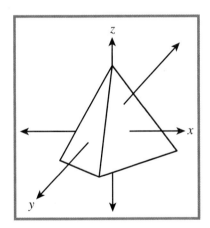

FIGURE 2.18 This object is rotating around the Z axis.

When you rotate the object front to back, you're rotating it around the X axis. And when you rotate the object left to right, you're rotating it around the Y axis.

All this talk of rotation around an axis can make you a little dizzy, I know. The good news is that I don't expect you to understand this concept of rotating around a specific axis completely at this point. You don't even need to have a super-solid grasp of planes and axes at this point. These concepts will start to make more sense when you dig in and start using a 3D modeling application like Tinkercad. So why don't we get started?

CHAPTER 2: 3D Modeling Basics

Up next in Chapter 3, "Say Hello to Tinkercad," you're going to get hands-on training with Tinkercad and learn how concepts such as axes, planes, and rotation (as well as many more) apply to it. I hope you're excited because once you start using a 3D modeling application, it just gets more and more fun, and you start finding new and interesting things to create. Let's go!

Say Hello to Tinkercad

In This Chapter
- Finding and opening Tinkercad
- Navigating Tinkercad
- Changing a 3D model's properties
- Looking at lessons

What kind of name is Tinkercad? What do these buttons do? What is this grid? Can I use a mouse with it? What about a touchpad? Can I use it on a tablet? Why not use software with more features?

The first two chapters are full of theory and concepts. In this chapter, you're now entering the hands-on section of the book. From this point forward, you're going to be getting some training in one of my favorite 3D modeling/CAD applications: Tinkercad.

Anyone who wants to learn to create digital 3D models is going to have to actually use the software. There's just no other way! You can read the chapters and look at the figures and nod to yourself that you understand what I'm attempting to explain, but nothing beats putting your hands on a mouse or touchpad and following along with the instructions and performing the tasks that I describe.

So sit yourself down in front of a computer or laptop and get ready for a quick tour of the software. You don't have Tinkercad installed? No worries. We'll get to that.

Rest assured that much of what you'll learn in this chapter will be applicable to other 3D modeling applications. Other software may have more or fewer menus, buttons, and toolbars, but if you get to know Tinkercad you'll be ready to jump into more advanced 3D modeling applications when you finish this book. (Or, like me, you may find that Tinkercad offers just about everything you need for your 3D modeling work.)

CHAPTER 3: Say Hello to Tinkercad

Finding and Opening Tinkercad

If a picture is worth a thousand words, Figure 3.1 should be worth at least a couple hundred. This is Tinkercad.

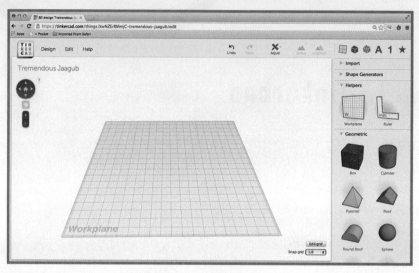

FIGURE 3.1 This is the Tinkercad application.

A few windows and options are hidden from view in Figure 3.1 (but can be easily shown by selecting certain buttons or menu options), but this view is 90% of Tinkercad. This is the user interface that you will be using throughout the remainder of the book.

Figure 3.1 shows Tinkercad running on my Apple MacBook Air, which runs the OS X operating system. You might be running Tinkercad on a Windows computer or laptop. If that's the case, most of Tinkercad will look identical, with the exception of the window controls that let you minimize, maximize, and close windows. On a Windows machine, the window controls are in the upper-right corner, whereas on a Mac they're in the upper-left corner.

I'll be going over all the various menus and buttons later in the book, but for now, I just want you to know where to find Tinkercad and how to access its features.

Tinkercad is not an application that you must install on your computer or laptop. Tinkercad runs in a web browser—the same tool you use to access websites like Google and Amazon. But Tinkercad won't run on just any web browser; it will only run on a browser that supports something called WebGL. What you need to know about WebGL is that it is software built into the browser that allows 3D objects to be rotated on your screen, among other functions. Three popular web browsers that will run Tinkercad are Firefox, Chrome, and some versions of Internet Explorer.

Finding and Opening Tinkercad

The best way to determine whether you can use Tinkercad is to open your web browser, point it to www.tinkercad.com, and click on the Start the Editor button.

If you get an error message like the one shown in Figure 3.2, you'll know you need to download and install a compatible web browser. (Visit firefox.com or chrome.com to download and install one of those web browsers.)

FIGURE 3.2 Unfortunately, not all browsers are WebGL compatible.

NOTE

If you are running Apple's Safari browser, you need to enable WebGL support on OS X and higher. Perform these steps:

1. In Safari, click the Safari menu and select Preferences.
2. Click the Advanced tab.
3. At the bottom of the window, place a check in the Show Develop Menu in Menu Bar checkbox.
4. Close the dialog box, click the Develop menu, and select Enable WebGL.

You can find a complete list of all operating systems that support WebGL by visiting http://caniuse.com/webgl.

For the remainder of this book, I use the Chrome web browser, but if you're using a different browser, there shouldn't be any major differences between what you see onscreen and what you see in this book's figures.

Once you have a browser that is compatible with Tinkercad installed on your computer, open it up and point it to www.tinkercad.com. You should see a welcome screen like the one in Figure 3.3. (The image may look different because the site sometimes changes the model features on the opening screen. No worries!)

FIGURE 3.3 Tinkercad's welcome page appears in the web browser window.

The first thing you're going to need to do is create a user account. Although you are not required to create an account to use it, Tinkercad offers both free accounts and paid accounts so that users can save their models online. Also keep in mind that one major difference between the two types of accounts is that free accounts don't offer the ability to work as a team on projects (with two or more people working on the same model at the same time), and you can't use your designs for commercial use. (In other words, you can't sell them.)

Click the Pricing menu shown in the upper-left corner of the screen in Figure 3.3 to see the various pricing plans and how a free account compares to a paid account.

Click the Sign Up for Free Account button and fill in the requested information (see Figure 3.4).

After you've created an account and signed in, you're taken to the standard Tinkercad Dashboard, shown in Figure 3.5.

When you first start using Tinkercad, you won't have any 3D models created, and thus you'll have no projects. You'll see only the free lessons.

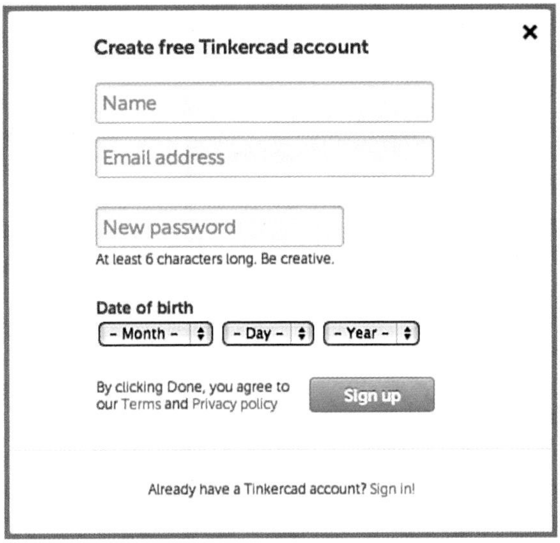

Finding and Opening Tinkercad

Create free Tinkercad account

Name

Email address

New password
At least 6 characters long. Be creative.

Date of birth
- Month - | - Day - | - Year -

By clicking Done, you agree to our Terms and Privacy policy   **Sign up**

Already have a Tinkercad account? Sign in!

FIGURE 3.4 Creating a free account with Tinkercad.

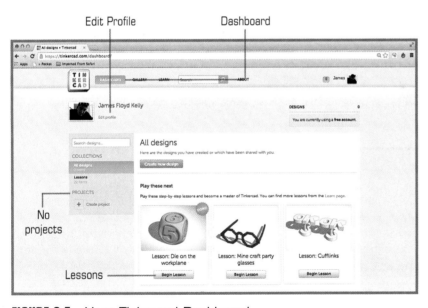

FIGURE 3.5 Your Tinkercad Dashboard.

CHAPTER 3: Say Hello to Tinkercad

You may have noticed that I have a picture associated with my account. If you'd like to add a photo of yourself (or some other image), click the Edit Profile button, and you see a screen like the one in Figure 3.6.

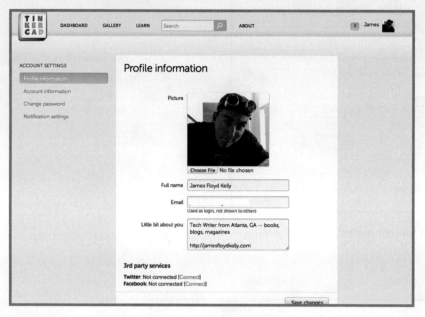

FIGURE 3.6 Your profile can contain an image, an email address, and a bio.

To return to the Tinkercad Dashboard, click the Save Changes button and then click the Dashboard button.

NOTE

If you are under 18, ask your parents or teacher if it's okay to add a photo to your Tinkercad account. For Internet safety reasons, it might be better for you to use the generic image provided by Tinkercad or include a favorite cartoon character or some other image that doesn't allow strangers to know what you look like. I also encourage you to be careful about what personal information you include in the Little Bit About You section and to not include your email address.

Now that you know how to access Tinkercad and have created an account, it's time to learn a few basics about navigating around the Dashboard and some of the menu options.

Navigating Tinkercad

The Tinkercad Dashboard isn't difficult to use. There's simply not a lot of options on it until you get into the actual design work of a 3D model. Figure 3.7 shows all the important areas that you'll be using occasionally.

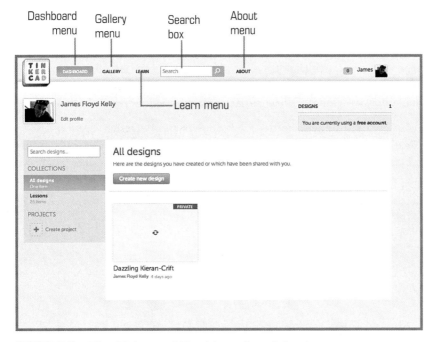

FIGURE 3.7 The Tinkercad Dashboard explained.

Here's a rundown of the areas in Figure 3.7:

- **Dashboard menu**—Click this menu to reach the basic Dashboard layout shown in Figure 3.7.
- **Gallery menu**—Click this menu to view and browse various 3D models designed by other Tinkercad users. You can filter them by using options such as Newest Things for most recent uploads, Staff Favorites for staff favorites, and Hot Now for 3D models that are really popular at the moment.
- **Learn menu**—Click this menu to view all the lessons that Tinkercad makes available to users. (You'll see where to find the lessons and how they work a bit later in this chapter.)
- **Search box**—If you know the name of a 3D model or the user who created it, type that in the search box and find it fast. Likewise, you can find 3D models by typing in words that you believe might be used to describe a model, such as *dice* or *robot* or *Eiffel Tower*. All the 3D models whose names fit your search term are displayed.

CHAPTER 3: Say Hello to Tinkercad

- **About menu**—Click this menu to get details about the web browsers and operating systems that are compatible with Tinkercad.
- **Edit Profile button**—Click this button to edit your profile or to log out.
- **Collections list**—You can click on the All Designs option in this category to view any 3D models you have created (or copied and modified from other user models). The Lessons option allows you to view all the lessons you have completed or are currently working on.
- **Projects list**—You can group 3D models you are designing into a project category. Models you create in a project collection show up when you select the All Designs option, but only 3D models created in a project can be shared with other users. (This feature is available only with a paid Tinkercad account, not with a free account.)

You'll be creating your own 3D models shortly, but for now, your All Designs section is probably empty. Click the All Designs option and then click the Create New Design button. Tinkercad opens a blank workspace. Click and hold one of the shapes to the right (such as the sphere) and drag it onto the workspace, as shown in Figure 3.8.

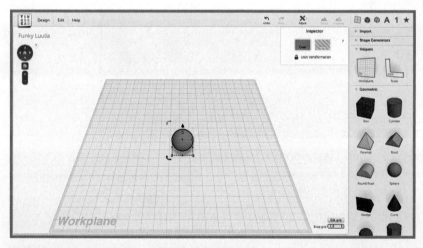

FIGURE 3.8 Creating a simple 3D model.

After you've dragged the shape onto the workspace, click the Design menu and select Save. Click the Design menu once more and click Close. Your simple 3D model now appears in the All Designs listing, as shown in Figure 3.9. It also will most likely have a very strange name!

Navigating Tinkercad

FIGURE 3.9 Your 3D model is now shown on the Dashboard.

> **NOTE**
>
> Tinkercad really has some fun naming the various 3D models you create, but you can easily change a name to something that's more useful to you. You'll learn how to do this in the next section.

Once your simple 3D model appears on the Dashboard, move your mouse pointer over the image. Do not click the Tinker This button that appears in Figure 3.10 but instead simply click the actual shape or model that appears.

CHAPTER 3: Say Hello to Tinkercad

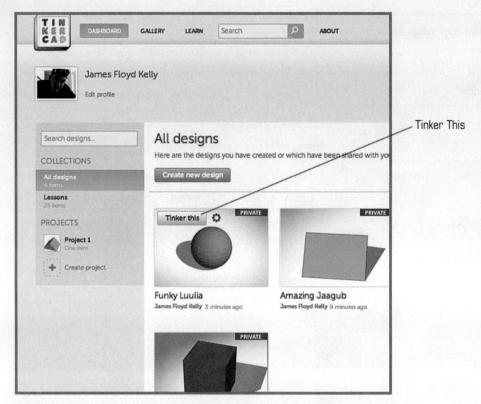

FIGURE 3.10 Click a model to open its Preview window.

The 3D model's Preview window opens, as shown in Figure 3.11.

You can click any 3D model that you find in the Gallery or in your own listings to open its Preview window. Using the Preview window, you can leave comments for the model's designer, click the Like button, share it on Facebook or Twitter, pin it to Pinterest, or click the big blue Tinker This button to open a copy in Tinkercad that you can modify any way you like. (This works only for 3D models that users have selected to share with the world. You'll more about this in Chapter 4, "Learn Some Modeling Basics.")

You may notice buttons that let you download a model for 3D printing, play with the model in the *Minecraft* game, or order a printed copy from a company that specializes in creating actual objects you can hold in your hand. I'll discuss these options later in the book, but for now, just know that you can access these options from the Preview window as well as the Design menu while you're creating a 3D model.

Finally, near the bottom of the Preview window are the View 3D button, the Standard view button, and the Add Photo button. Click the View 3D button, and you can rotate the object around on the screen and view it from different angles. The Standard view button (in the middle) shows you a view of the model as it was last saved, much like to a photograph. The Add Photo button lets you add additional images such as photos of an actual object you have created from the model or maybe a screenshot of the model from different angles.

One final item that you might have missed is in the upper-right corner of the Standard view or 3D view of the model. In Figure 3.11 you can see the word Private, which indicates that I've not yet elected to share this model with other Tinkercad users. I can change this later or keep the model private forever. Later in the book, I'll show you how to share your 3D models with other users, but for now, all 3D models you create in Tinkercad will initially be set to Private. (The alternative is Public.)

FIGURE 3.11 The Preview window for a 3D model.

Changing a 3D Model's Properties

Refer to Figure 3.10 and look to the right of the Tinker This button. See that small gear? Move your mouse pointer over it, and you see the Actions menu. Click it, and a drop-down list of options appears, as shown in Figure 3.12.

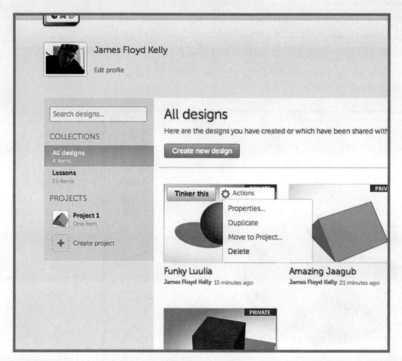

FIGURE 3.12 The Actions menu offers different options for 3D models.

The Delete option is fairly easy to figure out: Click it once, and you're asked to confirm that you're okay with deleting the model. Likewise, the Move to Project option lets you assign a 3D model to a particular project you are working on.

The Duplicate option creates an identical 3D model in your All Designs listing but with "Copy of" at the beginning of the model's name, as shown in Figure 3.13.

The Properties option opens a small window like the one in Figure 3.14. Here you can rename the 3D model by typing a new name in the Name field. You can click the Visibility drop-down menu to select Public if you want to share your 3D model with the world.

Changing a 3D Model's Properties

FIGURE 3.13 Easily copying models with the Duplicate option.

FIGURE 3.14 A 3D model's Properties window.

Figure 3.15 shows that I've renamed the copy of the 3D model named Copy of Funky Luulia and selected Public so anyone can download my wonderful blue sphere.

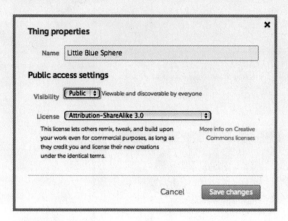

FIGURE 3.15 Using the Properties window to rename and share an object.

> **NOTE**
>
> The License drop-down menu offers a variety of permission types you can apply to allow others to copy and modify your 3D models—or not. Select each of the options to read a short description of what it allows or disallows and choose the one that you're most comfortable with. Free accounts are limited to the Attribution-ShareAlike 3.0 option, but paid accounts can apply the other options. Click the More Info on Creative Commons Licenses link for more detailed information on the various licensing options.

Click the Save Changes button to return to the Dashboard. Figure 3.16 shows that my 3D model now has the name Little Blue Sphere, and it is a Public model that you can see in the Gallery and that other Tinkercad users can download.

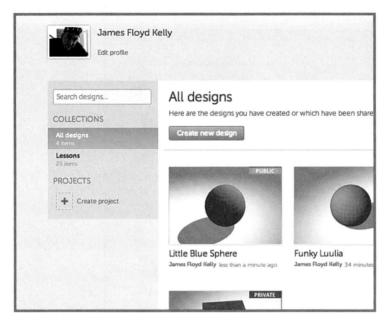

FIGURE 3.16 My new model is available to the world.

Looking at Lessons

Tinkercad comes with a number of free lessons that can teach you the basics of using many of the Tinkercad tools and menus. For now, though, you should probably hold off on working through those lessons. But if you'd like to see what's available in terms of lessons, simply click the Learn tab on the Dashboard, and you get a list like the one in Figure 3.17. Click on any lesson to open and start it. The lesson's onscreen instructions will tell you what to do.

The reason I'm hesitant to have you jump right into the lessons is that you could easily do them out of order, and that would be very frustrating. Over the next few chapters I go over many of the same tools and options that you'll learn in the lessons, and I provide more detailed explanations of how they work as well as additional hands-on examples that will help you grasp the fundamentals.

That said, if you'd like to try one or two or all of the lessons, feel free to do so and then come back to Chapter 4. But if you can hold off until after Chapter 9, "More Useful Tricks with Tinkercad," you'll be able to go back and fly through the lessons on your own and understand them a bit better.

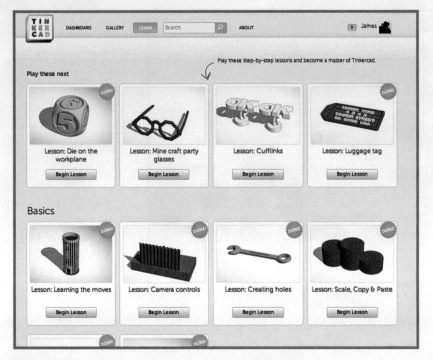

FIGURE 3.17 Sample lessons from the Tinkercad staff.

Here's what you've learned in this chapter:
- Use the Create New Design button to create a simple model
- Use the Design menu to save a model and then close it
- Use the Actions menu to duplicate a model or delete it
- Rename a file and set it to Public or Private
- Preview a model and add comments or share it with social media
- Find a lesson and open it

In Chapter 4, you'll begin creating unique 3D models and learn about the various toolbars and tools available to you. Let's get to it!

Learn Some Modeling Basics

In This Chapter
- The launchpad
- The rocket's main body
- The rocket's fins

It's time to start digging a bit deeper into the workings of Tinkercad, and the best way to learn about this amazing tool is to get hands-on with it. Now, before you can go and create your own amazing models, you've got to realize that there are a lot of basic skills you'll need to master first. This doesn't mean you can't create some 3D models right now, however. It just means that it might take you a little bit longer.

The better way to learn Tinkercad is to start slow and first learn how to use its most basic tools and features. An even better way to learn Tinkercad is to create an actual model as you're learning the ins and outs of the application. And that's exactly what this chapter is all about. By the end of this chapter, you'll have created all the components necessary for a simple 3D model and you will have learned many (but not all) of the standard features that Tinkercad offers. To show you how to make the parts of a simple 3D model, I've picked a fun little rocket for you to build. Go ahead and open up Tinkercad, log in, and click the Create New Design button on the Dashboard.

The Launchpad

This first model is simple. You'll be using basic shapes to create a model of a small toy rocket, preparing to launch. You'll need to create the pieces for both the launchpad and the rocket—starting with a launchpad that's sitting on a tiny piece of land. The first thing you're going to want to do is create that small piece of land. As you can see in Figure 4.1, I'm zoomed in quite a bit on the workspace. I need to zoom out a bit so I can see the entire workspace.

CHAPTER 4: Learn Some Modeling Basics

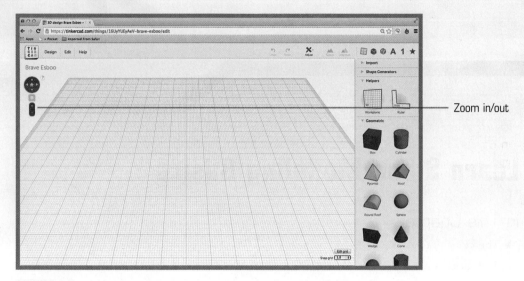

FIGURE 4.1 You can zoom in and out on the workspace.

To zoom in and out, you can use the + and – buttons indicated in Figure 4.1. A few taps on the – button shrinks the workspace a bit so you can see all of its boundaries, as shown in Figure 4.2.

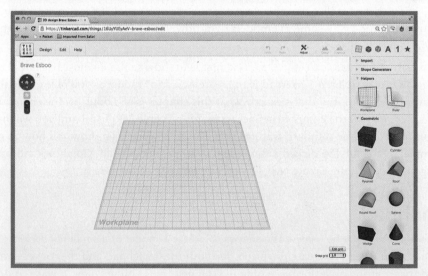

FIGURE 4.2 Zooming out to see the workspace boundaries.

Click the + button to zoom in and see more detail on a model. If you're using a mouse that has a mouse wheel on top, you can also scroll it away from you to zoom in and toward you to zoom out. Finally, if you're using a Mac touchpad, you can swipe two fingers down to zoom in and swipe two fingers up to zoom out.

Go ahead and drop a piece of the launchpad on the screen. To do this, you need to click, hold, and drag a copy of the red box onto the workspace. You can find the red box shape in the Geometric section of the toolbar that runs down the right side of the screen, as shown in Figure 4.3. If you don't see the red box, click the word Geometric to open the list of geometric shapes.

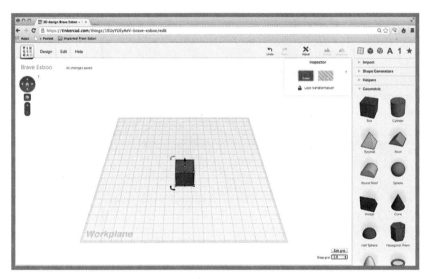

FIGURE 4.3 Dragging a red box object onto the workspace.

When you drop a shape on the workspace, it may or may not appear with various controls around it, such as arrows or tiny white boxes in the corners. When an object is selected, a few controls appear on and around it. In Figure 4.4, I've zoomed in on the box from Figure 4.3 so you can see these controls in more detail. If you don't see the controls on your screen, simply click the red box, and they appear.

You've dropped a box object onto the workspace, but it might not completely look like one from the angle shown in Figure 4.4. It would be nice to rotate the workplane a bit so it's a more obvious that this is a box. To do this, you can use the rotate controls shown in Figure 4.5. Likewise, if you're a mouse user, you can either press and hold both mouse buttons simultaneously while moving the mouse to see the workspace move or press and hold the middle (wheel) button; test both to see which works for you. Mac users can press and hold two fingers on a touchpad to achieve the same result.

CHAPTER 4: Learn Some Modeling Basics

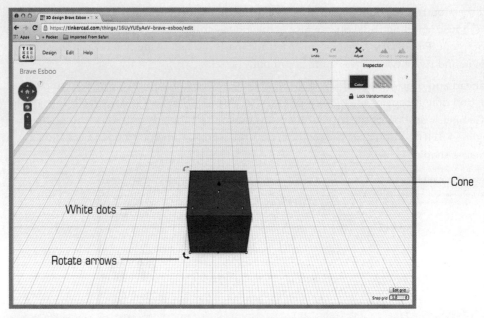

FIGURE 4.4 Controls allow you to manipulate an object.

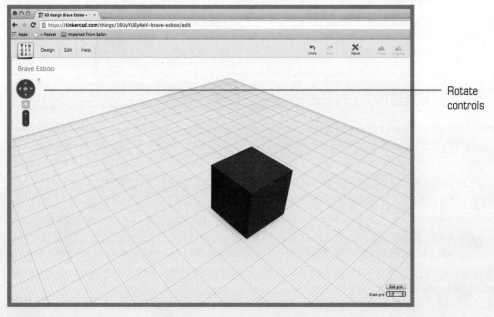

FIGURE 4.5 Using the rotate controls to change the view of the workspace.

The Launchpad

Find a suitable angle to view the box object and then click on the box object to select it so the controls are visible once again, as shown in Figure 4.6.

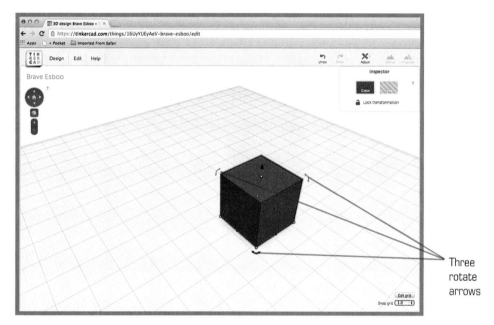

FIGURE 4.6 The controls on the selected box object.

Compare the controls shown in Figure 4.4 to those shown in Figure 4.6. With a simple shift of the workspace view, you should now see in Figure 4.6 that there are three rotate arrows surrounding the box instead of just the two shown in Figure 4.4.

You'll learn about the rotate arrow controls in Chapter 5, "Putting Together a Model," but for now I want you to focus on the small white dots that are visible in the bottom corners of the box object.

Move your mouse pointer over any white dot, and a measurement appears. Some dots, such as the one in Figure 4.7, displays two measurements.

As you can see in Figure 4.7, the length and width of this box object are both 20mm (millimeters). This means the base of this box is a square. You can check the height of the box by clicking the white dot control on the very top of the box: Move your mouse pointer over it as shown in Figure 4.8, and the height measurement appears.

Because the height is also 20mm, you're looking at a perfect cube. But a cube isn't the best place to launch a rocket. You're going to modify the cube so it's very flat and a bit larger, and you'll do it by clicking and dragging on those white dot controls.

CHAPTER 4: Learn Some Modeling Basics

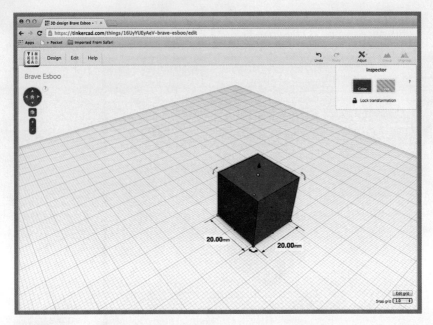

FIGURE 4.7 White dot controls display measurements.

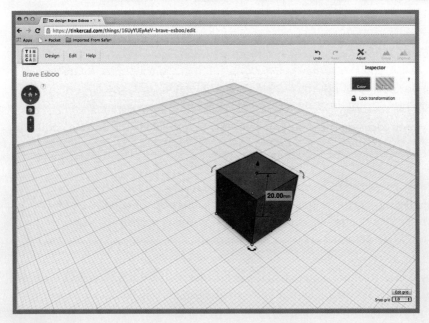

FIGURE 4.8 The top white dot control reports the height measurement.

Let's start with the height. Click and hold on the white dot on top of the cube while dragging down. Watch as the cube begins to flatten in size, and the height measurement value decreases. Notice also that the length and width remain the same: 20mm. Shrink the box object's height to 1mm and stop. Your box object should now look like the one in Figure 4.9.

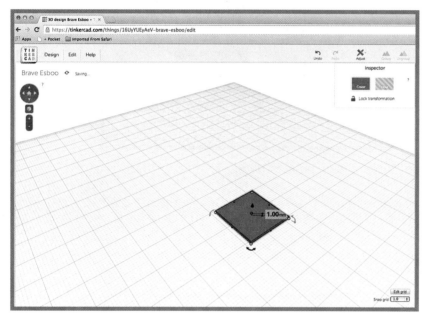

FIGURE 4.9 Flatten the box object by dragging down on the top white dot control.

You can verify the height at any time by hovering your mouse pointer over the top white dot control again. Once you're satisfied that the height is 1mm, click and drag on one of the white dot controls that make up the corners of the box object until the object's length and width values are both 100mm. When using one of the corner white dot controls, you can change both the length and width at the same time. Experiment a bit and see how moving the mouse pointer while clicking and holding down on a white box lets you change the length and width simultaneously.

Your box shape should end up looking like the one in Figure 4.10, at 100mm in both length and width and 1mm in height.

In Figure 4.10, the launchpad extends beyond the workspace. This isn't a problem, but it can affect your view when you zoom in and out of the workspace. Ideally, you want to try to keep your models within the boundaries of the workplane for easier viewing. To move an object such as the flattened launchpad, simply click and hold on any part of the object's surface (but not on any of its controls) and drag it and release it where you want it.

CHAPTER 4: Learn Some Modeling Basics

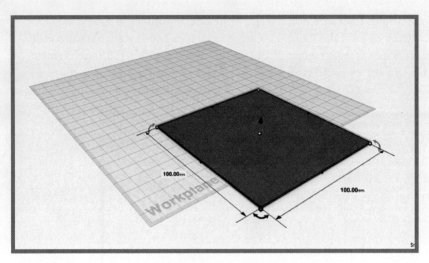

FIGURE 4.10 The rocket's launchpad is done.

This is a good time to point out just how easy it is to change the color of a selected object. When you have an object (such as the launchpad) selected, click the Color button indicated in Figure 4.11 and pick a new color. (You can use the Custom link to create a unique color if you like.) Figure 4.11 shows that I've changed the launchpad's color to green and centered the object in the middle of the workplane.

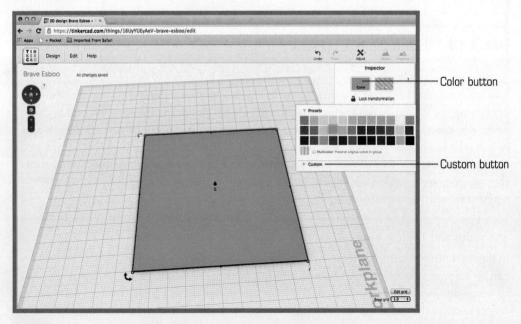

FIGURE 4.11 You can drag and drop model pieces anywhere on the workspace.

Before you move on to creating the pieces of the rocket, I want to give you some more practice with dropping objects on the workplane and modifying their sizes. To get this practice, you can create the launch scaffolding. This will consist of three pieces that will eventually be stacked. Just create four blue objects using the box shape object and give them these dimensions (length x width x height):

Scaffold1: 10mm x 10mm x 10mm

Scaffold2: 7mm x 7mm x 20mm

Scaffold3: 4mm x 4mm x 15mm

Scaffold4: 30mm x 5mm x 2mm

Drag and drop these launchpad scaffolding pieces around the edges of the workspace, as shown in Figure 4.12. You'll put these together in Chapter 5, along with the pieces of the rocket that you'll be creating next.

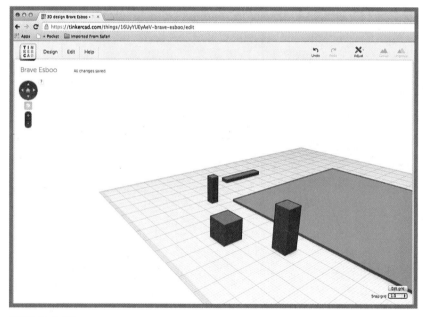

FIGURE 4.12 The scaffolding pieces, sized and ready for assembly.

The Rocket's Main Body

Don't worry, you're not going to be re-creating the Space Shuttle in Tinkercad...although that would make a great advanced project for you to consider. Instead, you're going to create the pieces for a small and simple rocket that will consist of the main body, the engine, and three fins.

CHAPTER 4: Learn Some Modeling Basics

When the scaffolding is assembled, the height of the assembly will be 47mm. The rocket will have a height of 45mm. Keep this in mind as you create the main body, engine, and fins.

It can often be useful to have a hand sketch of your model (or a photo of a real-life object), and Figure 4.13 gives you the basic idea of the rocket you'll work on creating. Your version doesn't have to look exactly like mine, but do try and follow along with the hands-on steps so you get some practice with the Tinkercad tools and features used here.

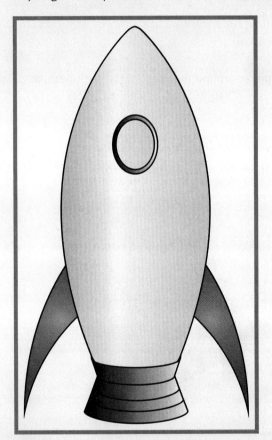

FIGURE 4.13 A sketch of the rocket 3D model to be designed.

As you can see, the largest part of the rocket will be the main body, and I'm going to start with that shape. The easiest way to obtain this shape is to drag and drop a Sphere object on the workspace and use the white dot controls to modify the shape. Figure 4.14 shows that I've dropped a sphere object onto the workspace.

Like a box, a sphere also has the white dot controls appear when you select it. Unlike a box object, however, a sphere's measurements are related to its diameter (when it's a perfect sphere). You can click and hold on any white dot and manipulate the shape of the sphere as desired. After some playing around, I managed to end up with an elongated cigar-shaped object, as shown in Figure 4.15, with length and width of 14mm and height of 30mm.

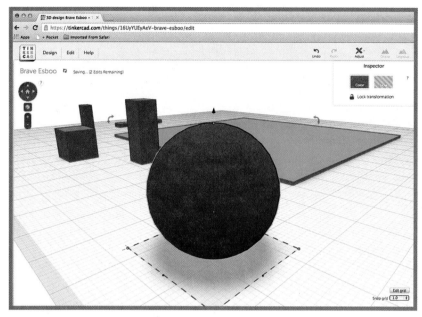

FIGURE 4.14 A sphere object, ready to be modified.

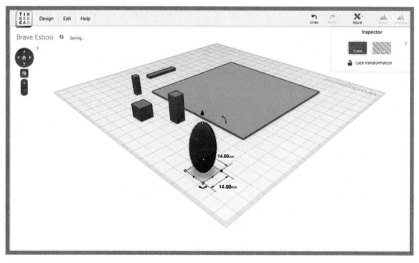

FIGURE 4.15 The rocket's main body is cigar shaped.

CHAPTER 4: Learn Some Modeling Basics

> **NOTE**
>
> You may be wondering if you need to save your work at this point. Don't worry. Tinkercad is constantly saving your progress. However, to save your progress manually, you can click the Design menu and then click the Save button.

Now let's move on to the engine. Creating this part involves simply sizing a small ring that will be placed on the bottom of the main body. You'll learn in Chapter 5 how to properly line it up and center it on the main body. For now just drag and drop a tube object (which you find at the bottom of the Geometric section) onto the workspace and give it a diameter of 8mm and a height of 2mm.

Because you want to keep the diameter of the ring constant, the length and width must be the same value. If you hold the Shift key down on your keyboard as you drag one of the white dots inward, the length and width are locked together and shrink at the same rate. Figure 4.16 shows the completed ring.

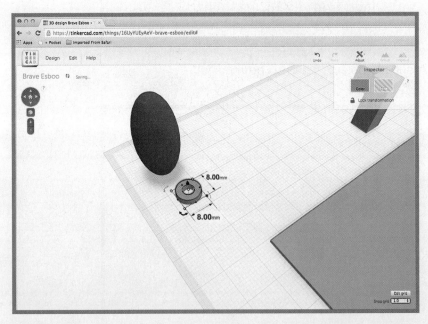

FIGURE 4.16 The engine ring, resting on the workspace.

The Rocket's Fins

Now it's time for the fins. For the fins, you'll need to use an interesting little trick to create the curves, but once you've made one basic fin, you'll be able to make two exact copies of it to save time. The trick to the fins involves creating what's called a hole object and using it like a cookie cutter to remove unwanted areas of a solid object.

You'll find as you continue to work with Tinkercad that the application sometimes requires you to be a bit creative in order to get the shapes you want. As you can see by browsing the Geometric section of the toolbar, there is no fin shape. But that's not going to stop you. You can first focus on the outer curve of the fin shown in the sketch in Figure 4.17.

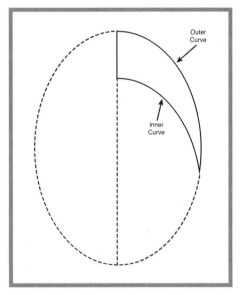

FIGURE 4.17 A fin will be a small sliver of a larger object.

As you can see, the outer edge of the fin is actually a piece cut out of a stretched circle. You'll first create a flattened cigar-shaped oval that has the curve you want for the outer edge. You'll do this by dropping a thick tube object on the workplane, as shown in Figure 4.18.

CHAPTER 4: Learn Some Modeling Basics

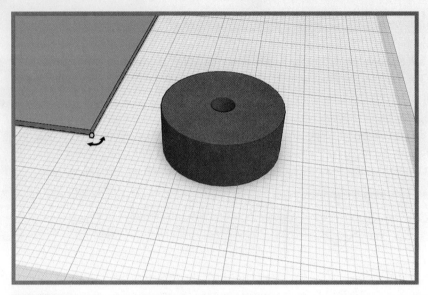

FIGURE 4.18 Start with a thick tube object.

Flatten the tube to 2mm in height and stretch the tube to get the curve you want. Figure 4.19 shows that I've flattened the tube to 2mm and dragged one of the other white dot controls to give the finished oval dimensions of 18mm wide by 35mm long.

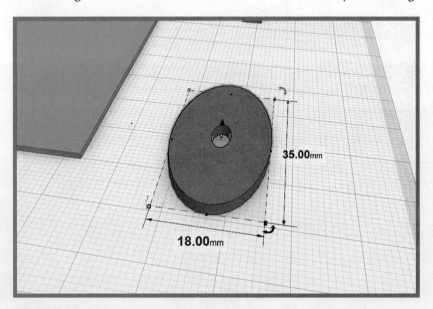

FIGURE 4.19 The flattened oval is the start of a fin.

Next, cut the tube object in half vertically by creating a 2mm-thick rectangle to cover half of the tube. Figure 4.20 shows that I've dropped a box object onto the workspace, shrunk it to 2mm in height, and matched its width and length to that of the tube object. (The width is 18mm and the length is 35mm.)

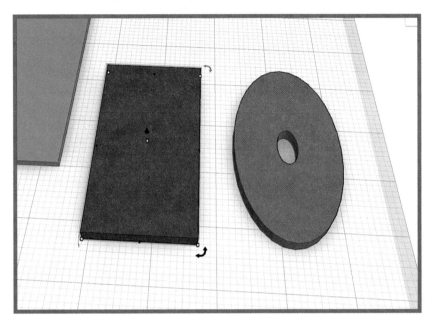

FIGURE 4.20 The flat rectangle will help cut the oval in half.

Now, here's the trick: Select the rectangle object and then click on the Hole button indicated in Figure 4.21. Notice that the rectangle will change from a color to a clear outline.

Now, drag the rectangle so that it covers half of the oval object. Figure 4.22 shows the rectangle object overlapped over the oval.

CHAPTER 4: Learn Some Modeling Basics

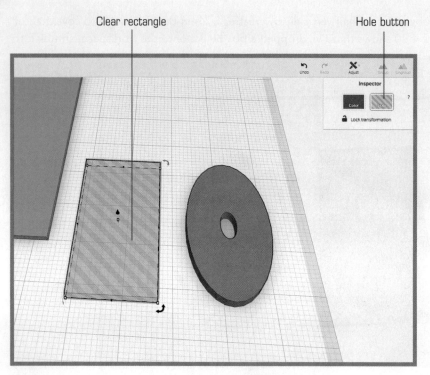

FIGURE 4.21 Turn the rectangle into a hole object.

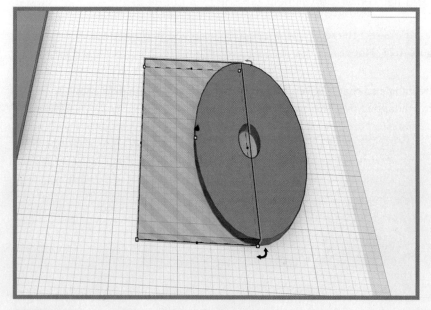

FIGURE 4.22 Overlap a solid object with a hole object.

Finally, drag and select both objects so they are outlined. Likewise, you can hold down the Shift key and click on each object to perform a multiple-select action. Figure 4.23 shows that the clear hole object and the orange oval are selected.

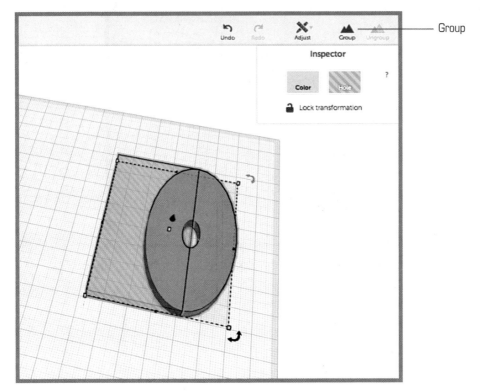

FIGURE 4.23 Select both objects to be combined.

Click the Group button. This instructs Tinkercad to combine the two objects into a single object (see Figure 4.24).

As you can see, the outer hole objects will be treated as empty space and will delete or erase any solid objects they encounter. This means that half of the oval object will disappear. Keep in mind that Tinkercad must perform some calculations to determine where solid material and a hole intersect, and this can sometimes take a minute or two for complex combinations. When the calculations are done, the final piece that is left exists as a single object, as shown in Figure 4.25.

CHAPTER 4: Learn Some Modeling Basics

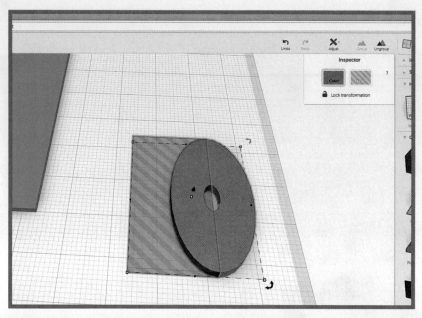

FIGURE 4.24 Group two or more objects together to create a single object.

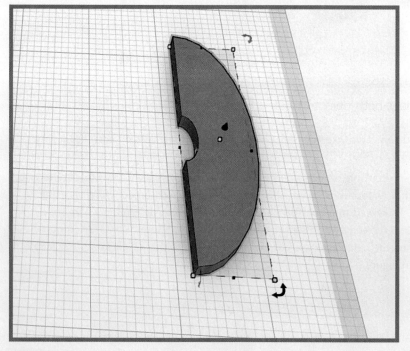

FIGURE 4.25 The final object consists of material not combined with the hole object.

The Rocket's Fins

NOTE

While the calculations for the grouping of the two objects is being done by Tinkercad, you might think that nothing is happening and hit the Undo button in haste. Be patient. Sometimes a merge can take a few seconds and other times a few minutes. Rest assured: If you clicked the Group button, Tinkercad is busily trying to figure out what to keep and what to delete. The flipside is that a click of the Undo button can also take some time to undo the grouping. If the grouping took more than a minute to complete, expect the Undo button to take about the same time.

Now take a look at Figure 4.26, and you'll get an idea of how the final shape of the fin will be accomplished.

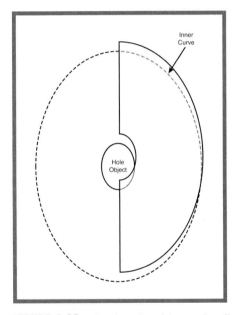

FIGURE 4.26 A sketch of how the final fin will be made.

CHAPTER 4: Learn Some Modeling Basics

> **NOTE**
>
> The fun thing about the Group button and the Hole button is that you can keep using them over and over, refining the shape of an object as you slowly delete away bits and pieces by merging a hole object with a solid object. One suggestion is to change the color of an object before turning it into a hole object. While both objects are solid, the distinct colors will help you distinguish between each object as you merge them. Once you're happy with the merge, you can convert one object to a hole object—you don't have to convert it to a hole object before the two objects are merged.

To make the inner curve of the fin, you need to create another unique object on the workplane and then convert it to a hole object. Then you will group the new hole object with the half oval in Figure 4.25, which will yield the final desired fin shape. The dimensions of this second oval piece are 49mm x 64mm. Figure 4.27 shows a new oval object, flattened to 2mm and shaped to get the desired inner curve.

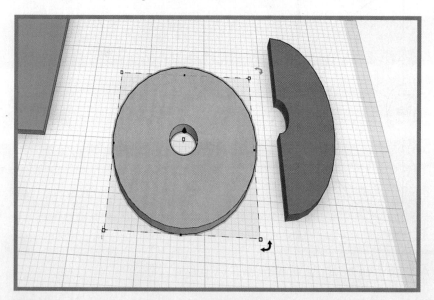

FIGURE 4.27 A fin will be a small sliver of a larger object.

After selecting the new shape and clicking the Hole button, drag the two objects together, as shown in Figure 4.28. You can see that the final shape of the fin will be the bit that's not covered by the hole object.

When you're happy with the fin shape, select both objects and click the Group button again. The hole object disappears, along with any solid sections of the oval object, leaving only the final fin shape, as shown in Figure 4.29.

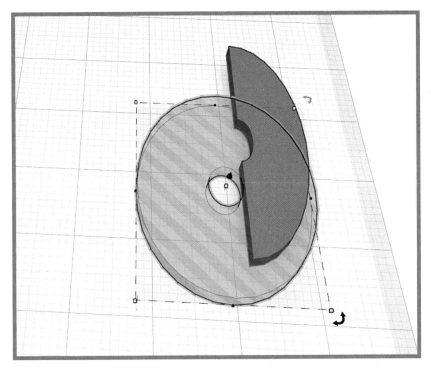

FIGURE 4.28 Creating the final fin requires another hole object.

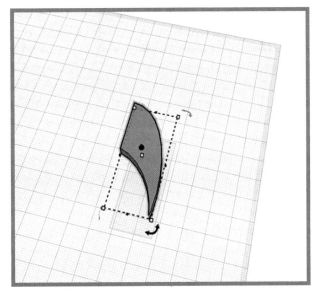

FIGURE 4.29 The final fin shape.

Now all that's left is to create two more fins. The fin should already be selected because when you group two objects, the final object is always selected when the grouping task is complete. You can simply make a copy of the object by pressing Ctrl+C on a Windows computer or Command+C on a Mac. You can also click on the Edit menu, as shown in Figure 4.30, and then click the Copy option.

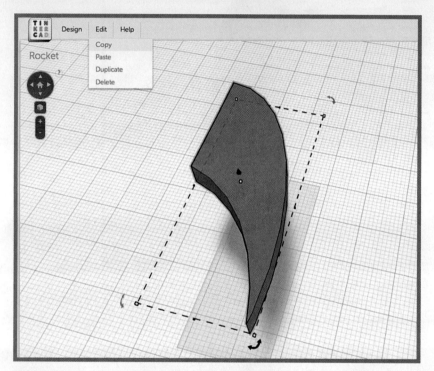

FIGURE 4.30 Copy the fin object and make two more.

Select the Edit menu again and click the Paste option to have a copy of the fin object placed on the workspace. Instead of using the menu, Windows users can press the Ctrl+V shortcut, and Mac users can press Command+V.

A pasted copy usually overlaps the original by a small amount. Just click the copy and drag it to a blank area of the workplane. Figure 4.31 shows my three final fin objects.

Zoom out a bit, and you can see all your parts ready for assembly to make the rocket and the launchpad, as shown in Figure 4.32.

Up next in Chapter 5, you'll learn how to combine the parts. This will involve some stacking and centering of objects as well as some rotating of parts (such as the fins).

The Rocket's Fins

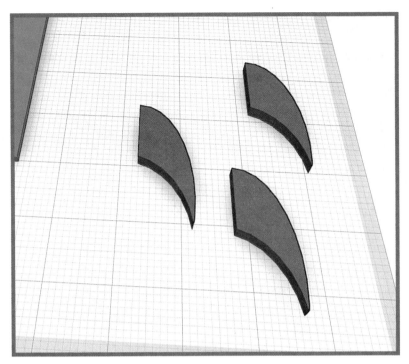

FIGURE 4.31 Three fins, ready to be attached to the rocket.

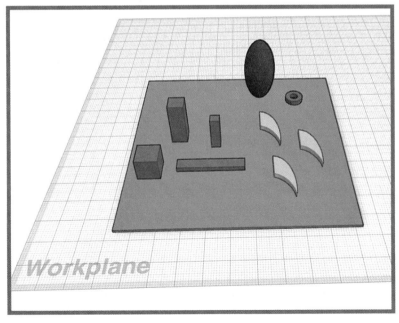

FIGURE 4.32 All the objects are ready for assembly.

CHAPTER 4: Learn Some Modeling Basics

Before you leave this chapter, think about some of the new skills you have acquired in Tinkercad:

- **Dragging an object around on the screen**—This involves simply clicking and holding on an object and placing it where you want it on the workplane.
- **Zooming in and out**—This will be useful when you want to get closer to a particular piece (such as one of the fins) for detail work or to view all your work at once.
- **Resizing objects**—By using the white dot controls, you learned how to modify an object's width, length, and height.
- **Changing color**—Changing the color of an object can make distinguishing parts of the larger model easier. Once parts are combined, a single color can be selected.
- **Converting an object to a hole object**—By creating an object and turning it into a hole object, you can remove material from a solid object.
- **Selecting multiple parts**—You can select more than one part at a time by holding down the Shift key and picking them one by one or else dragging a rectangle around them.
- **Copy and paste**—Copying and pasting copies of an object will save you lots of time when you begin making more complicated models.

You'll be learning many more Tinkercad skills in the chapters to come, but for now, feel free to create a new design of your own and play around with the other shapes available in the Geometric section. Take some time to experiment and use the skills you've acquired so you'll feel confident using them throughout the remaining chapters. See you in Chapter 5.

5

Putting Together a Model

In This Chapter
- Assembling the launchpad
- Assembling the rocket

While you were creating the individual pieces that will make up the launchpad and rocket, you were getting practice in using the most basic tools that Tinkercad offers. You learned to drop shapes on the workspace and change their dimensions easily with the white dot controls. You also learned the simple skill of creating an object and turning it into a hole object, which you can then combine with a solid object to remove portions and leave behind a completely new shape. In addition, you learned about copying and pasting, which will save you a lot of time and frustration as you begin to put together more advanced models.

The rocket model you'll be finishing up in this chapter is certainly not advanced. It's made up of simple shapes and a few unique ones (the fins) made using a hole object. But I'm hoping you're beginning to get a better idea of just how far you can push Tinkercad when it comes to making your own models. You'll have to frequently get creative in order to make the shapes you want for your models, but with patience and some thought, you can typically find a way.

In this chapter, you're going to start assembling all those rocket and launchpad pieces. As you do this, you'll learn some additional features that Tinkercad offers. Once again, I highly encourage you to actually perform the work and follow along as I show you how this is done. By performing the actions yourself, you'll be reinforcing your book knowledge and moving it to your long-term memory by doing physical movements such as using menus, clicking buttons, and dragging of the mouse pointer.

When you're done with this chapter, you'll have a solid grasp of almost all the key features and tools Tinkercad offers. You'll start with the assembly of the launchpad to learn the basics of stacking parts and lining them up properly, two skills that are necessary for putting together any 3D model, simple or advanced.

Assembling the Launchpad

Take a look at Figure 5.1, and you'll see the four pieces that make up the launchpad. You will shift them around and stack them to create the scaffolding that will sit to the side of the final assembled rocket.

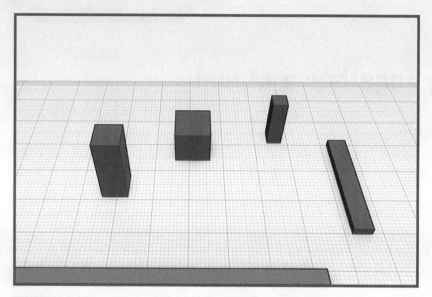

FIGURE 5.1 Four pieces of the launchpad scaffolding.

When all the pieces are assembled, the final 3D model will look like the one in Figure 5.2. Notice that the fins are no longer lying flat on the workplane but have been rotated and joined to the main body. You'll learn how to make that happen later, but right now you need to get the scaffolding built.

The base piece will be the 10mm × 10mm × 10mm cube. Sitting on top of the cube will be the 7mm × 7mm × 20mm piece, followed by the 4mm × 4mm × 15mm block. These three pieces will be centered on the Z axis that runs right down the center of each block. Only the final piece, the 30mm × 5mm × 2mm block, will be placed so that it sticks out and over the top of the rocket's nose cone.

Right now it doesn't matter where you assembled the scaffolding, but you will want to drag each of the four pieces so they are visible on the screen and not overlapping any other pieces. Use the Zoom In and Zoom Out feature to get the four pieces visible on the screen.

The base piece (10mm cube) will stay put; there's no need to move it or rotate it. Instead, you'll be focusing on the 7mm × 7mm × 20mm piece. You need to lift this piece and place it on top of the 10mm cube. So select the 7mm × 7mm × 20mm piece, as shown in Figure 5.3.

Assembling the Launchpad

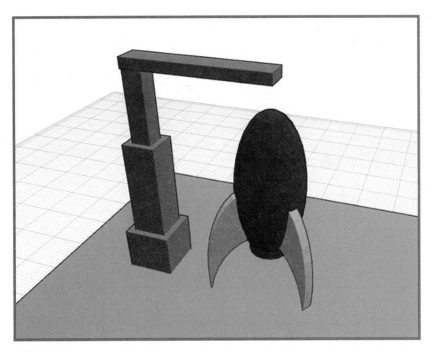

FIGURE 5.2 The final 3D model, with scaffolding sitting to the side of the rocket.

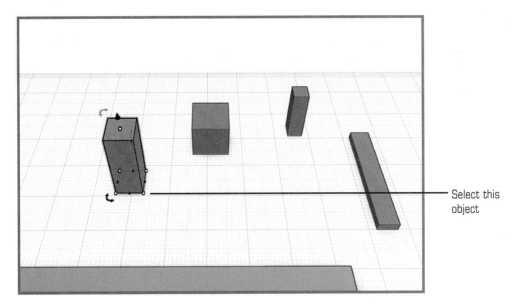

Select this object

FIGURE 5.3 Select the first piece of scaffolding to be stacked.

CHAPTER 5: Putting Together a Model

When you select this piece, you see the familiar white dot controls that let you resize the piece. In this case, though, you want to select the small black cone that's just above the white dot control you would use to increase or decrease the height of the block. This is the Raise/Lower control.

Click and hold the Raise/Lower control and drag your mouse pointer up. You'll see that the height above the workspace is indicated as the counter increases (see Figure 5.4). You know that the 10mm cube is 10mm in height, so it should be obvious that you need to raise this 7mm × 7mm × 20mm cube exactly 10mm above the workspace before sliding it over and placing it on top of the 10mm cube.

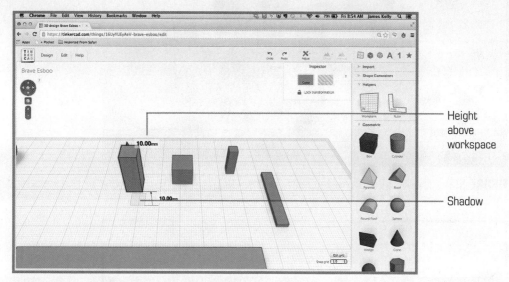

FIGURE 5.4 Raise an object above the workspace.

After you've raised the object 10mm above the workspace, release the Raise/Lower control, and the piece remains where you placed it. A shadow of the object's base remains on the workspace, indicating where the piece would sit if you lowered it back down to the workspace.

If you need to move a piece back down to the workspace and have forgotten how high (or low) it sits with respect to the workspace, simply move your mouse pointer over the Raise/Lower control, and a small measurement appears, as shown in Figure 5.5, telling you the distance from the workspace.

This piece is sitting 10mm above the workspace, and you need to move it so that it's sitting on top of the 10mm cube. You can do this by clicking any colored part of the shape and dragging it so it sits on top of the cube.

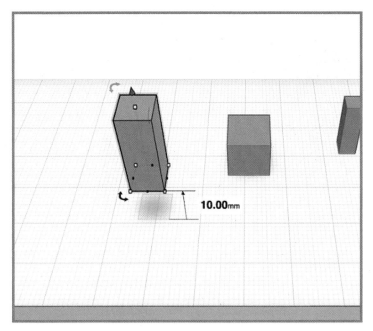

FIGURE 5.5 An object's distance from the workspace is visible.

Figure 5.6 shows that I've placed one box on top of the other, but they're not quite centered. While you want the two pieces to be centered along the Z axis, you don't have to worry about placing the top piece exactly in the center of the bottom piece; Tinkercad has a tool for doing just that.

To get these two boxes centered, you use the Align feature. To understand how the Align feature works, select the 10mm cube that forms the base of the launchpad by clicking it once. Then click the Adjust button, as shown in Figure 5.7, and select the Align option.

CHAPTER 5: Putting Together a Model

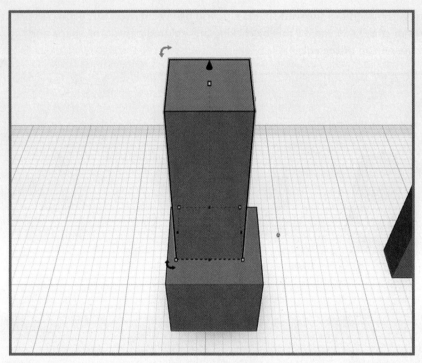

FIGURE 5.6 The first two boxes are now stacked but not centered.

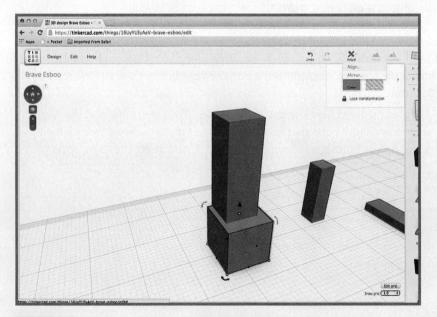

FIGURE 5.7 Select a single object and view its Align options.

After you click the Align option, you should notice nine gray dots appear. Three of them run parallel to the length of the object, three run parallel to the width, and another three run parallel to the height. You can see all nine dots in Figure 5.8.

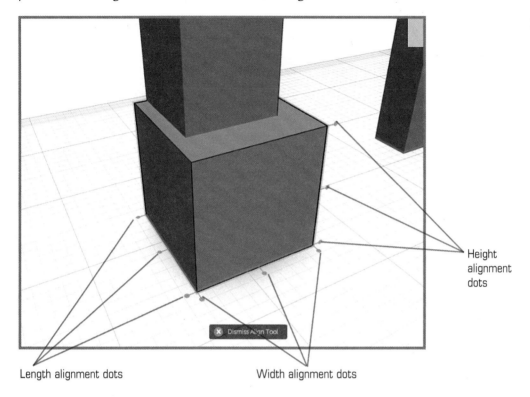

FIGURE 5.8 The gray dots help align parts.

Let's focus on the width alignment dots in Figure 5.8 for just a moment. Notice that one dot connects to the front face of the cube with a small line that matches up exactly to the front face/edge (which is resting on the workspace). Another dot connects to the back face of the cube with another small line, but this one matches up exactly with the rear edge of the cube. Finally, the third dot is exactly halfway between the front and rear edges of the cube and has its own small line connecting the alignment dot to the cube.

The best way to understand how these alignment dots work is to actually use them, so right now hold down the Shift key and select both the 10mm cube and the block sitting on top of it. (Likewise, you can drag a selection window around both objects with your mouse pointer to select them both at once.)

CHAPTER 5: Putting Together a Model

After selecting both objects, click the Adjust button and select the Align option once again. Figure 5.9 shows that once again, the nine alignment dot controls appear.

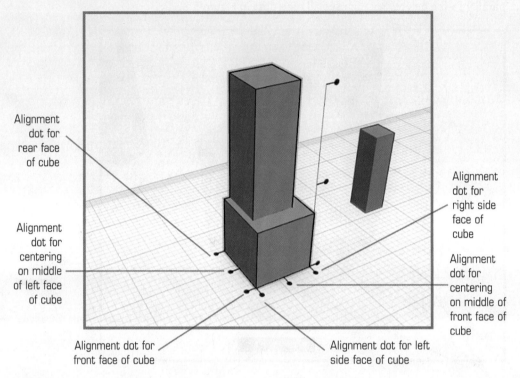

FIGURE 5.9 The Align feature works for multiple selected objects.

Once again, let's focus on the alignment dot controls running down the length of the object. Move your mouse pointer over the length alignment dot closest to the front of the 10mm cube—but don't click it yet. Notice that a yellow preview outline of the box on top appears and that it has been shifted so its front edge matches the 10mm cube front edge in Figure 5.10.

If you actually click that alignment dot, the cube on top shifts its position so both objects are aligned along that front edge. Figure 5.11 shows what this looks like.

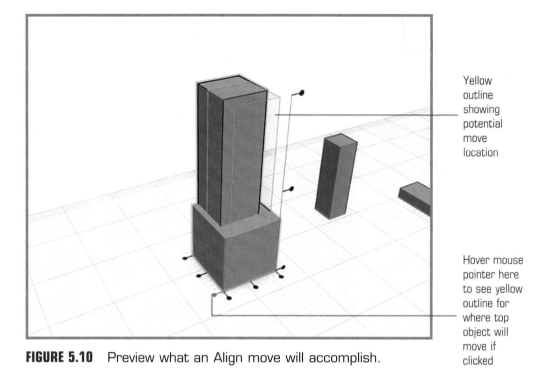

FIGURE 5.10 Preview what an Align move will accomplish.

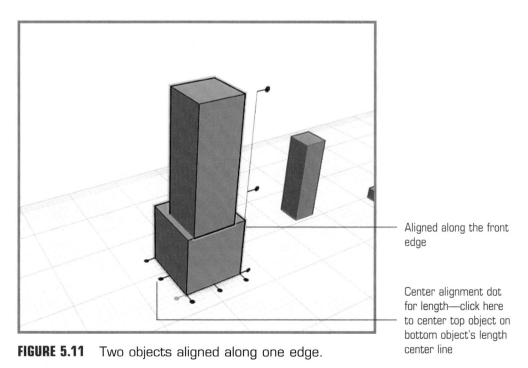

FIGURE 5.11 Two objects aligned along one edge.

CHAPTER 5: Putting Together a Model

If you click the alignment dot in the back (the one that indicates the rear edge of the 10mm cube), the top piece moves back so that the two objects now have their rear edges aligned.

But it's the center alignment dot that you really need here. Make sure both objects are still selected and the Align feature is active and click the center alignment dot. Figure 5.12 shows that both objects are now centered on that line with respect to the objects' lengths.

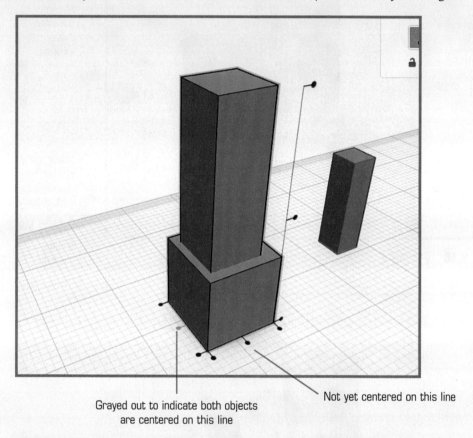

Grayed out to indicate both objects are centered on this line

Not yet centered on this line

FIGURE 5.12 Both pieces are centered along their lengths.

You're not done yet. You still need to center both objects with respect to their widths as well. Look closely at Figure 5.13, and you can see that while both objects are centered along their lengths, they are not centered along their widths. I've rotated the view a small amount so this is more obvious.

Assembling the Launchpad

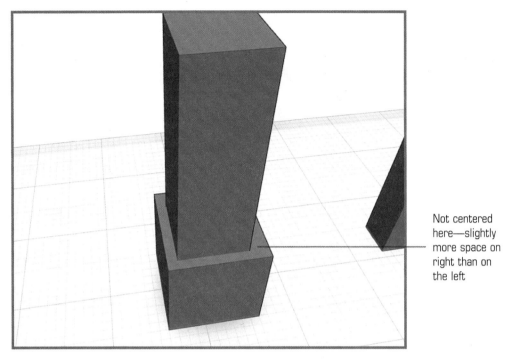

FIGURE 5.13 The two objects need to be centered along their widths.

With both objects selected and the Align option active, you now should focus on the alignment dots running parallel to the width of the 10mm cube, as shown in Figure 5.14.

If you click the left or right alignment dots, the top block shifts so that it shares either a left edge or right edge with the 10mm cube. But it's that center alignment dots that you really need to use. Move your mouse pointer over the center alignment dots (but don't click), as shown in Figure 5.15, and you see the yellow outline that shows you where the top box will rest if you click the center alignment dots.

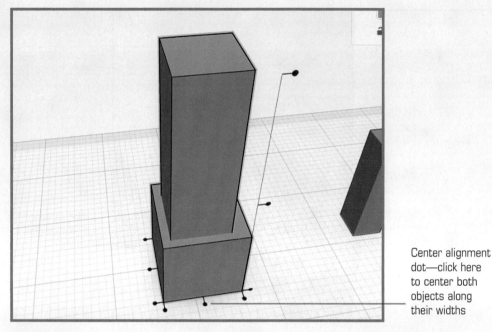

FIGURE 5.14 The alignment dot for centering along the width of the objects.

Center alignment dot—click here to center both objects along their widths

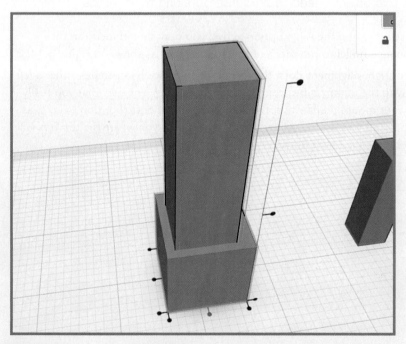

FIGURE 5.15 The center alignment dots moves the top box here.

Go ahead and click that center alignment dots, and the 7mm × 7mm × 20mm box shifts slightly, centering itself with respect to the 10mm cube's width. Figure 5.16 shows that the top box is now centered perfectly, and both boxes are centered on the imaginary line running vertically right down their centers (that is, the Z axis). You can see a hint of the Z axis line running down through the center of both boxes.

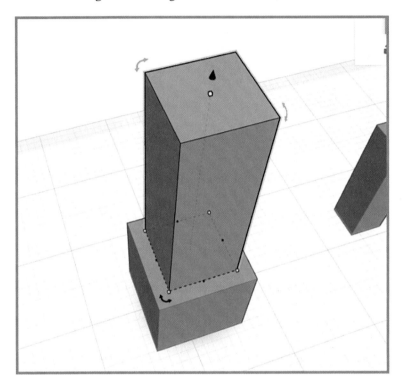

FIGURE 5.16 The two boxes are now centered.

At this point, you need to stack the remaining two boxes to finish the scaffolding. Raise the next box (4mm × 4mm × 15mm) and place it on top of the 7mm × 7mm × 20mm box; then center it on the Z axis. To determine how far above the workspace to raise this item, simply select both the 7mm × 7mm × 20mm box and the 10mm cube and then hover your mouse pointer over the white dot on top of the assembly. You could also add the heights of the two boxes, but this way, you can be certain. Figure 5.17 shows that the current height is 30mm.

Use the Raise/Lower control for the third box and raise it up 30mm. Figure 5.18 shows that I've raised it and placed it on top of the existing assembly. You need to use the Align feature again to center this box.

CHAPTER 5: Putting Together a Model

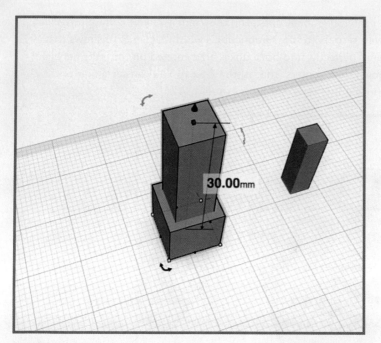

FIGURE 5.17 Find the height of an assembly consisting of multiple objects.

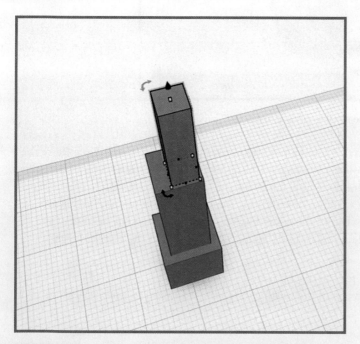

FIGURE 5.18 The third box is placed but not centered.

Assembling the Launchpad

You can choose to select only the top two boxes before using the Align tool, or you can select all three boxes. (Remember that the middle box is centered on the bottom box, so all you really need to do at this point is center the top box with respect to the middle box.)

> **TIP**
>
> If you get in a habit of selecting all the objects (using the Shift key), you can maintain that all objects are centered or aligned as you tinker with the Align tool. While you typically need to select only two objects at a time to use the Align tool, in this example three of the four objects are all centered down the Z axis, so selecting all three while using the Align tool helps make certain that none of the three base parts shift unexpectedly.

Use the Align tool once again, along with the center alignment dots, to center the top box with respect to the middle box's length and width. Figure 5.19 shows that all three boxes are now centered along a shared Z axis.

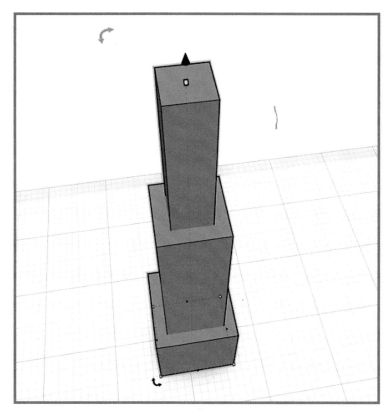

FIGURE 5.19 Three boxes centered along the Z axis.

All that's left to finish the launchpad scaffolding is to raise the fourth box to the top of the assembly. You need to raise it 45mm, and you don't need to center it.

In the case of this fourth and final box, you want to center it on its rear side with respect to *only* the box beneath it. For this reason, you should select only the top two boxes at this point. If you select all four objects and use the center alignment dots, you'll see something like the object shown in Figure 5.20.

FIGURE 5.20 The top scaffolding piece is centered.

But this isn't what the final assembly should look like. The top box needs to have its rear edge aligned with the rear edge of the 4mm × 4mm × 15mm box beneath it. Drag the top object so that it is as close to its final desired position as possible but still leave a small amount of the piece below it as visible, shown in Figure 5.21.

I've rotated the view to look at the back of the launchpad, and I've selected the top two objects and the Align tool. Notice that you can click on the indicated alignment dot so that these two boxes will share their rear edges. The yellow outline is subtle, but if you look carefully, you can see that the preview shows that the top piece will be shifted back a small amount.

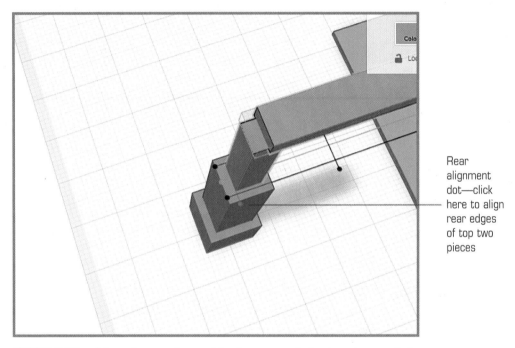

Rear alignment dot—click here to align rear edges of top two pieces

FIGURE 5.21 Align the top box with the rear of the box beneath it.

Figure 5.22 shows the final launchpad assembly, with all four boxes moved onto the flat green object. The rocket's main body, engine ring, and fins are all sitting around the perimeter, waiting to be assembled.

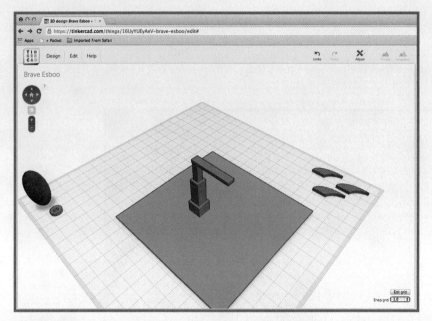

FIGURE 5.22   The final launchpad assembly.

Assembling the Rocket

The rocket consists of five pieces: main body, engine ring, and three fins. Start by attaching the engine ring to the main body; the engine ring must be centered along the Z axis, on a line that will run straight down the center of the main body. There are two ways to do this:

- Raise the ring above the main body and use the Align tool to center it before moving it down to its final position.
- Raise the main body and use the Align tool to center it over the engine ring.

In this case, use the second method: Raise the main body about 10mm above the ring, as shown in Figure 5.23.

You can see in Figure 5.23 that the shadow of the main body is visible on the workspace, and the engine ring is almost (but not perfectly) centered. Rather than refer to the ring and main body's length and width, we'll refer to the axes (X and Y) for the moment.

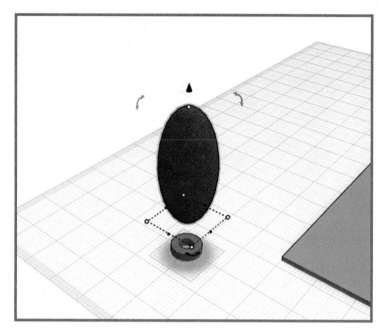

FIGURE 5.23 Select both objects to be combined.

Remember that the Z axis is typically used to indicate an object's height. Because you can rotate the workspace around, the width and length of an object can appear to change, depending on the angle from which you're viewing the object. Rather than worry about length and width, which are imprecise terms, you can center both objects along the X and Y axis, with the X axis running left and right and the Y axis running forward and backward with respect to the current view. In Figure 5.24, I've used the center alignment dots to align both objects on the X and Y axes. Notice that both alignment dots for centering the object on the X and Y axes are grayed out to indicate they are in effect.

Once that the main body and engine ring are centered on the Z axis, you can raise the ring and merge it with the main body. As you use the Raise/Lower control, the ring may completely disappear in the main body, remaining invisible, as shown in Figure 5.25.

CHAPTER 5: Putting Together a Model

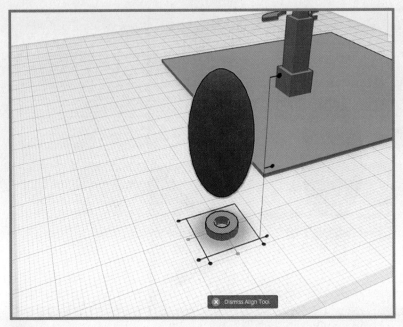

FIGURE 5.24 Align the ring and main body on both the X and Y axes.

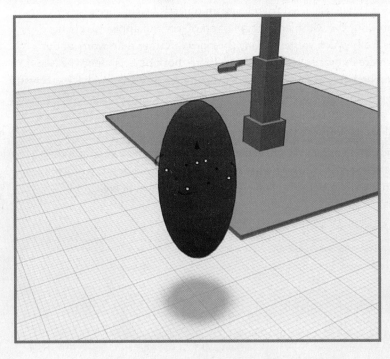

FIGURE 5.25 The ring is contained completely in the main body of the rocket and is not visible.

If you continue to hold down the Raise/Lower control, however, the ring eventually reappears, at the top of the main body, as shown in Figure 5.26.

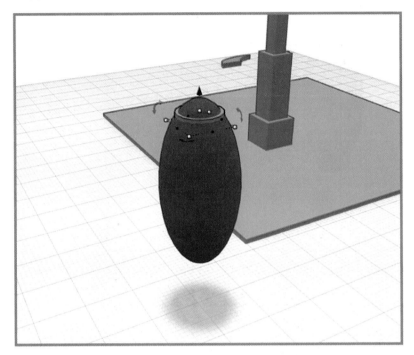

FIGURE 5.26 The engine ring has moved completely through the main body.

The engine ring should be placed near the bottom of the main body, and Figure 5.27 shows one possible placement for it.

Now it's time for the fins, which are currently lying flat on the workspace (see Figure 5.28). The three fins should circle the main body. This means you'll need to place the fins 120 degrees apart (because 360 degrees divided by 3 is 120). But before you place the fins in a circular pattern around the rocket's main body, you need to change their current position.

Right now, you need to manipulate a single fin to make it "stand up." Then you'll do the same thing with the other two fins.

NOTE

You could instead make one fin stand up and then make two copies of it, but if you follow the process described in this section, you'll get some hands-on experience (practice) using some important new tools.

CHAPTER 5: Putting Together a Model

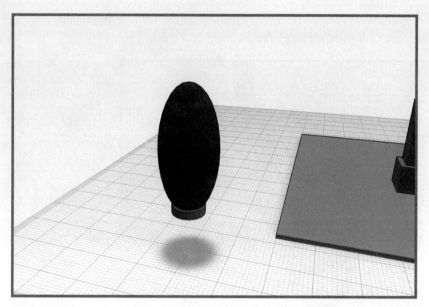

FIGURE 5.27 The engine ring in its final location on the main body.

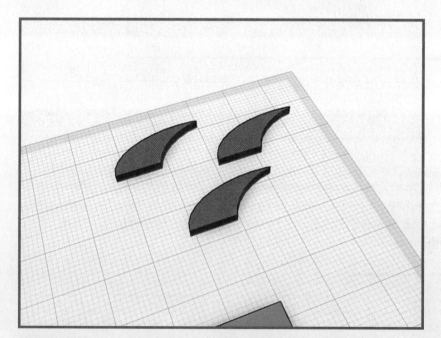

FIGURE 5.28 The three fins are lying flat on the workspace.

Click one fin to select it and take note of the small curved arrows that surround it (see Figure 5.29).

Assembling the Rocket

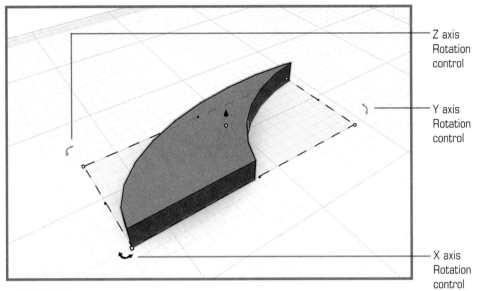

FIGURE 5.29 Select one fin that will be rotated.

Think back to Chapter 2, "3D Modeling Basics," where you learned about rotating an object along its X, Y, and Z axes, and what you're doing here will make sense. When you select an object, Rotation controls allow you to rotate the object along one axis.

For example, click and hold on the Z axis Rotation control and move your mouse around. Notice that the object spins in place as if an imaginary pin were holding it in place in the object's exact center point. There's even a measurement (in degrees) for precision rotating. Figure 5.30 shows the fin still lying flat on the workspace.

To make the fin "stand up," you need to rotate it along one of the other axes. Can you guess along which axis and how many degrees a fin must be rotated to make it stand up? If you said the X axis and 90 degrees, you're correct. Figure 5.31 shows that I've selected the X axis Rotation control and moved the mouse until the degree measurement equals -90. (Rotated the other direction, it would be +90 degrees, and the fin would be standing up but upside down.)

TIP

There are three rotation controls, one per axis, but sometimes they can be a bit difficult to see. Most often, one of the rotation controls is hard to see because of your view. If you find a rotation control hard to spot, simply rotate the object a bit left or right, and the missing control will often appear.

CHAPTER 5: Putting Together a Model

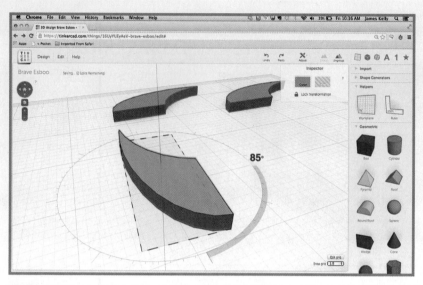

FIGURE 5.30 A fin rotated around its Z axis.

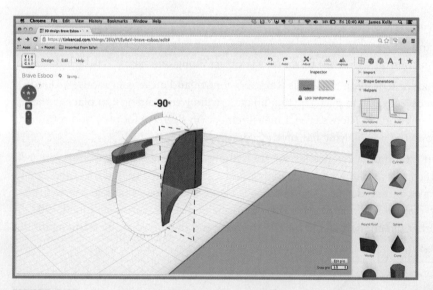

FIGURE 5.31 The rotated fin is now standing up.

Assembling the Rocket

> **TIP**
>
> If you hold down the Shift key as you rotate an object, the object jumps in 45-degree increments rather than 1-degree increments.
>
> Also, by having the cursor further away from the rotation tool, you gain greater accuracy in rotating objects than when the cursor is close to the tool.

Look carefully at Figure 5.32, and you'll notice that the fin's color changes slightly near its middle. This is because as the fin was rotated along the X axis, part of it dropped "beneath" the workspace surface. The lighter color indicates that part of the object is lying under the work surface, and you already know how to fix this: by using the Raise/Lower control.

FIGURE 5.32 Part of the fin is beneath the workspace.

Raise the fin until its bottom tip is resting on the workspace. If you hover the mouse pointer over the Raise/Lower tool, you see measurement that tells how far above the workspace the object is. Figure 5.33 shows that in this case, you need to raise the object 9.86mm.

If there's a decimal/fraction amount, don't worry: Just raise the object enough, and it should "snap" a tiny amount, to 0.00mm, as shown in Figure 5.34.

CHAPTER 5: Putting Together a Model

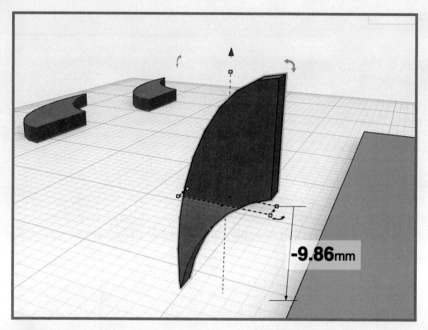

FIGURE 5.33 Raise the fin so it's completely above the workspace.

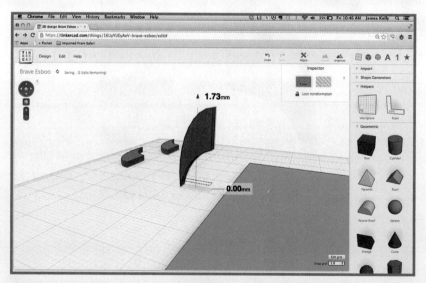

FIGURE 5.34 The fin is now resting on the workspace but standing up.

Perform this rotation on the remaining two fins, and you'll have three fins standing up side-by-side and ready for placement, as shown in Figure 5.35.

Assembling the Rocket

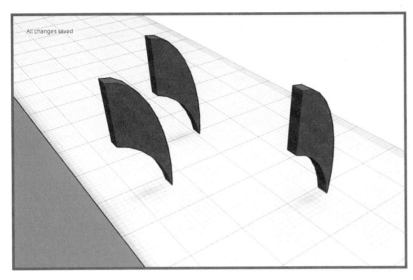

FIGURE 5.35 The three fins are now resting on the workspace, standing up.

Now you need to rotate the three fins so that they are each 120 degrees apart. This may seem like a complicated task, but it's really not. You already know how to rotate an object around its Z axis, and that's all you do here: Just rotate one fin around the Z axis 120 degrees and then rotate another fin around the Z axis 240 degrees. (You'll leave one fin as it is, representing the 0-degree position.)

Figure 5.36 shows that I've selected one fin and rotated it using the Z axis Rotation control until it's showing 120 degrees.

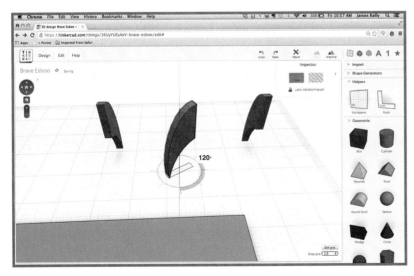

FIGURE 5.36 Rotate one fin 120 degrees.

CHAPTER 5: Putting Together a Model

> **NOTE**
>
> If you make a subtle mistake rotating the fin, you can always click the Undo button (to the left of the Adjust button) to go back one step. Multiple clicks on the Undo button will take you back additional steps, so use it carefully. You can also use the Redo button if you clicked the Undo button accidentally; when you do, whatever action was undone is redone.

The Rotation tool only counts up to 180 degrees before flipping to negative (–) measurements. One easy way to fix this is to simply rotate the object in the opposite direction so that you see the degree measurements in negative numbers. In this case, you rotate the third fin –120 degrees to get it to its final position. Figure 5.37 shows the third fin, rotated to –120 degrees.

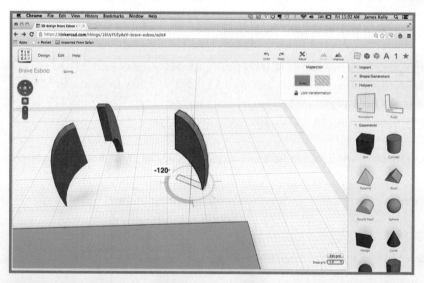

FIGURE 5.37 Rotate the last fin to –120 degrees.

Now all that's left is to drag and drop the fins onto the rocket's main body. Figure 5.38 shows the fins grouped closely together.

Now for the final trick. Not only do I want the fins to be the same color, but I'd prefer to be able to move them around the workspace as a single object rather than having to select them using the Shift key or by dragging a selection window around them. To group the fins, you need to select all three of them and click the Group button, as shown in Figure 5.39.

Assembling the Rocket

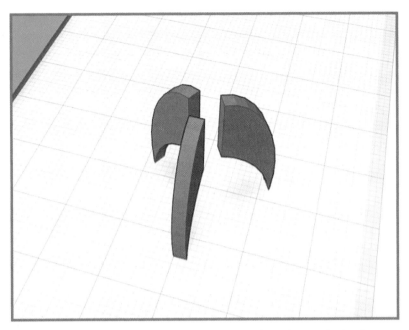

FIGURE 5.38 Group the fins prior to moving the main body.

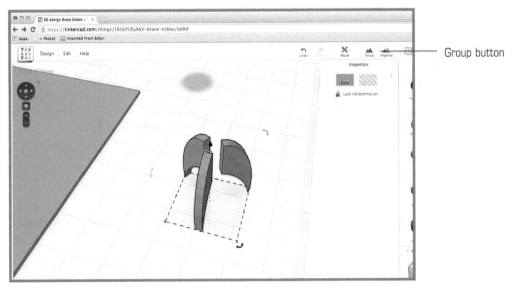

FIGURE 5.39 Make the three fins work as a single object by grouping them.

CHAPTER 5: Putting Together a Model

Any objects that are selected when you use the Group button are combined together and then act as a single object. For example, you can click the Color button to change the color of all three fins at the same time. You can also click any colored part of a grouped object and move the entire collection around. Try it!

Select the main body and the engine ring and group those two objects as well. Now you can drag the main body/engine ring as one piece and move it over the fins, trying out different heights until the fins are just where you want them.

You can do one final grouping of the fins with the main body/engine ring, but if you do, you'll discover one limitation of the Group feature: All grouped items must have the same color. If you want to have the fins remain a unique color, different from the main body, you should not use the Group feature to join the main body with the fins. In that case, if you want to move the fins and the main body/engine ring assembly together, you'll have to use your mouse to draw a selection box around all the parts before you can move the entire rocket. Select the rocket and move it under the launchpad assembly, as shown in Figure 5.40.

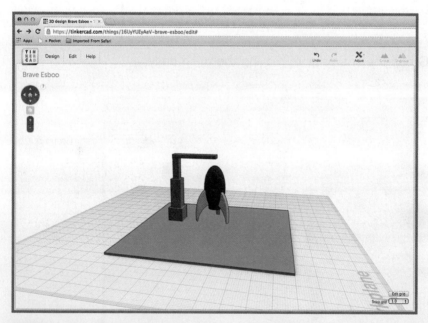

FIGURE 5.40 The final rocket in place and ready for launch.

You could perform some additional tweaks to enhance your new rocket model. You already have the skills to perform these tasks, but you might have to do some thinking to figure out exactly how to do a few of them:

- Increase the size of the rocket. By grouping the rocket (or the fins and the main body/engine ring assembly), you can enlarge or shrink the collection by using the white dot controls.
- Add a viewing window to the rocket. You can resize a ring object or a sphere object for a window, and you can use the Rotation controls to place it on the main body in such a way that it follows the main body's contour/shape.
- Add some numbers on a fin. Scroll down below the Geometric Shapes section on the toolbar, and you'll find the Letters and the Numbers sections, which contain pre-created letters and numbers that you can resize and rotate as desired.

You've learned a lot in this chapter! These are some of the new skills you've acquired in this chapter:

- **Grouping items**—By using the Group feature, you can select multiple objects and have them behave as a single object for purposes of assigning a color, moving the grouped object, and resizing and rotating.
- **Rotating on an axis**—Using the Rotation controls, you can rotate an object (or a group of objects) on the X, Y, or Z axis. You can rotate in 1-degree increments or hold down the Shift key to rotate in 45-degree increments.
- **Aligning objects**—You learned how to use the Align feature to force objects to share an edge or a center line. This is useful for centering objects with respect to one another or for forcing objects to line up a similar side or face.
- **Using Undo and Redo**—By clicking the Undo (or Redo) button, you can roll back changes (or redo them) that you've made to your models.
- **Adjusting the distance of an object above or below the workspace**—By using the Raise/Lower control, you can move a selected object up or down with respect to the flat workspace.

With these basic Tinkercad skills under your belt, you're ready to try designing a 3D model from scratch. You'll start with a hand sketch idea and put all your new Tinkercad skills to work to create something fun and functional. That's right—a 3D model that you can have turned into a real object to hold and share.

6

A Tinkercad Special Project

In This Chapter
- Brainstorming ideas
- Creating the basic tag shape
- Adding embellishments
- Adding raised text
- Suggestions for improvements

Now that you've got a grasp of the basic tools and features of Tinkercad, it's time to put them to work. Tinkercad is fun for creating 3D models like the rocket you've make in the past couple chapters, but it's also fun for designing things that can actually be put to use. Industrial designers use CAD applications (with much more advanced features than Tinkercad) to design functional items. If you've recently purchased something that was mass produced, you can bet that it was initially designed on a computer screen with a CAD application.

CAD applications allow you to quickly and easily make changes to your designs. Imagine creating prototypes by hand from materials like clay, plastic, and metal. Not only is that old-fashioned design process time-consuming, but it can get expensive, depending on the materials needed to create those initial designs. With CAD, you can finalize your design—the look, the measurements, and even moving parts—before the object becomes a physical item that can be touched.

Often a design begins as nothing more than a hand sketch on a piece of paper. And that's where you're going to start this chapter's special project—with an idea and a sketch or two. Then you'll move to Tinkercad to create a digital model of your sketch.

CHAPTER 6: A Tinkercad Special Project

> **NOTE**
>
> You may at some point want to convert a digital model into a real-life object. You'll learn more about that in Chapter 8, "Printing Your 3D Models."

If you like this project, feel free to tweak it, change it, and twist it to suit your needs. Or, if you've got something else in mind, run with that! Start with a sketch and then take it to Tinkercad and see what you can do.

Brainstorming Ideas

My son's school lets the kids run on the track on Wednesday mornings before school starts. They are presented tiny tokens that they can hang on their book bags to commemorate certain key tasks, such as running 10 miles, running 25 miles, and running the most laps in a month. These kinds of rewards help inspire the kids to show up (especially on cold mornings) and exercise, and it occurred to me that these token rewards could be expanded to cover many more achievements.

All members of the U.S. military are provided with a set of ID tags (also known as dog tags) that contain relevant personal information (such as blood type). They're worn on a chain around the neck so they don't get lost. This chapter builds on the idea of creating something similar to dog tags that could be used to reward your friends or family members for certain achievements. You'll create an interesting shape to be used over and over, and you'll be able to create copies of it to customize with unique text.

Figure 6.1 shows a traditional dog tag shape, along with some other ideas for other shapes to consider. Note that each one has a small hole for inserting a string, leather, or metal chain so it can be worn around the neck or tied to something.

I prefer the traditional dog tag shape over the circle shape. The star would be fun to create, but there's not a lot of room to add text in its body. And while I like the arrow shapes, the sharp end could be dangerous if worn.

The great thing about Tinkercad is that you can use it to create as many shapes as you like, and the only extra cost is your time. You can use the steps in this chapter and create as many basic shapes as you like. For now, you're going to create the traditional dog tag shape and add some fun embellishments. You'll save the basic dog shape as a blank; later, whenever you like, you can copy it and add text to it.

What kinds of embellishments can you add to a simple dog tag? Take a look at Figure 6.2, and you'll see some sketches of a couple ideas for that shape.

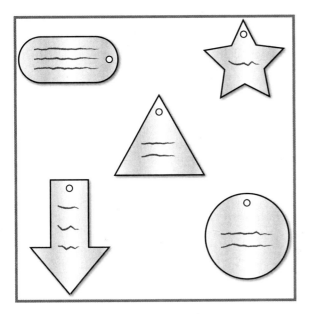

FIGURE 6.1 Brainstorming ideas for dog tag shapes.

FIGURE 6.2 Brainstorming ideas to make the tag more unique.

Here are two possible ideas:

- On the left and right sides of the tag, add a shape (such as a star) that will actually be a hole in the tag. Letters could be substituted in place of a shape.
- Raise both the outer edge of the tag and any text that goes in the center.

CHAPTER 6: A Tinkercad Special Project

How about some sort of combination of these two ideas? Remember, Tinkercad allows you to play around with your models as much as you like. You can create the basic shape—a rectangle with rounded edges on the left and right sides—and then make copies of it to test the various embellishments.

Creating the Basic Tag Shape

As you've probably noticed in the previous chapters, Tinkercad uses metric measurements by default. But you can easily change this if you like. For this project, change the unit of measurement to inches instead of millimeters. To do this, open a new project, as shown in Figure 6.3, and click the Edit Grid button.

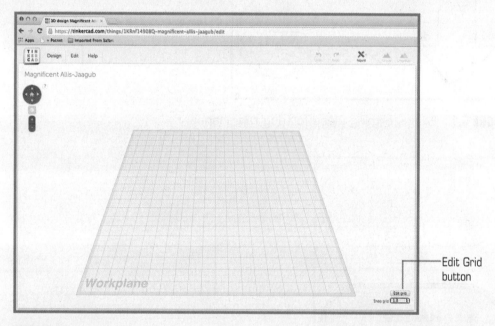

FIGURE 6.3 Use the Edit Grid button to change the units of measure in Tinkercad.

The Grid Properties dialog appears, as shown in Figure 6.4. It allows you to select either millimeters or inches for the units. You can also decrease or increase the size of the workspace grid if you like. Click the Update Grid button to close the Grid Properties dialog and return to the project.

Creating the Basic Tag Shape

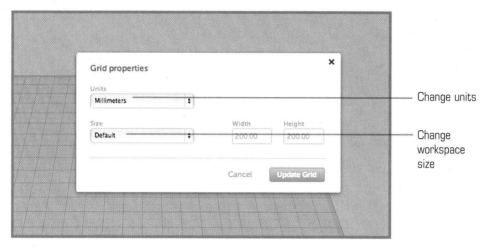

FIGURE 6.4 Change the units to inches and modify the workspace size.

TIP

You'll learn in Chapter 8 how 3D printers can be used to print out your 3D models in plastic. You'll also learn that each 3D printer has its own unique limits on the size of objects it can print. By changing the dimensions of the workspace in Tinkercad to match the dimensions of a 3D printer, you can always make certain the length and width of your 3D models doesn't exceed the print capabilities of the 3D printer. You'll learn more details in Chapter 8.

For this particular tag, I want the dimensions to be 2.5in × 1.5in × 1/8in (width × length × height), but you can change them to suit your own needs.

If you look below the Edit Grid button, you see another useful tool: Snap Grid. Click this button, and you see a drop-down menu appear like the one in Figure 6.5.

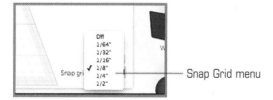

FIGURE 6.5 The Snap Grid tool helps you line up items and size objects.

Experiment with Snap Grid to find the setting that works best for you. The workplane grid shown in Figure 6.5 consists of 1/8in. squares and 1in. larger grids. When you have the Snap Grid feature set to 1/8in., any dragging or resizing of objects automatically jump to the nearest small box. Because each tiny square is 1/8in., dragging controls such as the white dot controls forces them to jump in 1/8in. increments and to always land on the edges of the 1/8in. squares. Figure 6.6 shows that as you resize the box object, the white dot controls jump from 1/8in. to 1/4in. to 3/8in. as you increase the width.

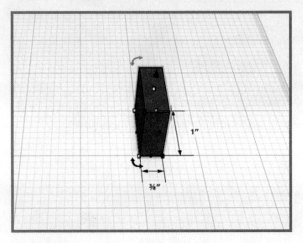

FIGURE 6.6 Objects that you move or resize jump in set increments.

Now see what happens when you change the Snap Grid setting to 1/16in. Figure 6.7 shows that as you increase the width of the box object, it now jumps in increments of 1/16in., and the width increases from 3/8in. to 7/16in. or 0.437in.

For purposes of your dog tags, setting Snap Grid to 1/8in. should be sufficient. But if you want finer control over resizing and creation of objects, decrease the Snap Grid setting to a small increment, such as 1/16in. or even 1/32in.

Now it's time to create the basic dog tag shape. The easiest place to start is with a simple rectangle. The final dimensions of the dog tag will be 2.5in. × 1.5in. × 1/8in. Because the edges of the tag will be rounded, you need to do some simple math to figure out how large to make the initial rectangle. The length of the tag will be 1.5in., and this means that the rounded edges on the left and right will consist of circles with diameters of 1.5in. Half of each circle is 0.75in., so if the overall width of the dog tag is to be 2.5in., you need to subtract each half of the circle (0.75in. × 2), which leaves you with a rectangle 1in. wide. Figure 6.8 shows how you figure out these measurements.

Creating the Basic Tag Shape

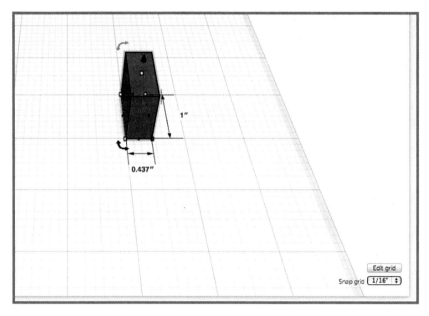

FIGURE 6.7 Change the Snap Grid setting to alter the incremental jumps.

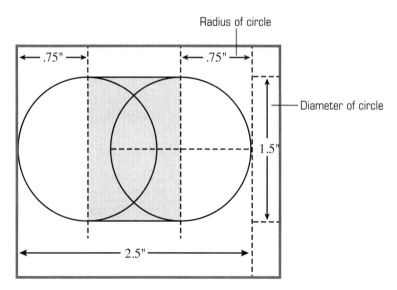

FIGURE 6.8 Calculate the width of the shaded rectangle needed.

Now place a rectangle on the workspace. It should have dimensions of 1in. × 1.5in. × 1/8in., as shown in Figure 6.9.

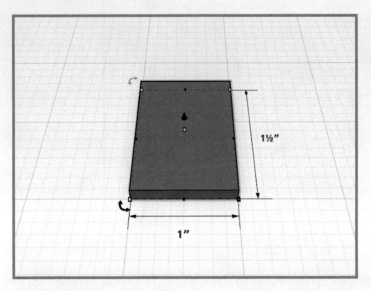

FIGURE 6.9 The initial rectangle is placed and sized appropriately.

Now it's time to add the curved edges. To do this, drag and drop a cylinder object onto the workspace, as shown in Figure 6.10.

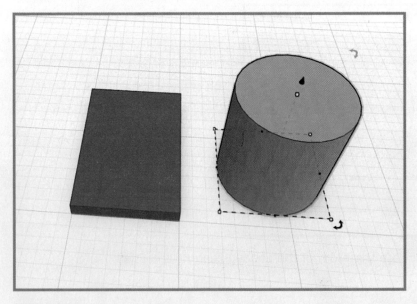

FIGURE 6.10 Use a cylinder to create the curved edges.

Resize this cylinder so that it has a height of 1/8in. Use the white dot control on top to drag the cylinder's height down to 1/8in. and then release. Figure 6.11 shows the flattened cylinder.

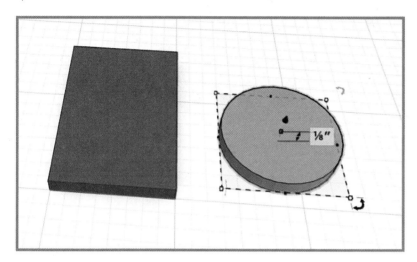

FIGURE 6.11 Flatten the cylinder to the proper height.

The cylinder object is a perfect circle, but you can use the white dot controls to easily change it to an oval if your length and width values are different. Because in this case you want the cylinder to have a perfectly circular shape, hold down the Shift key while dragging one of the length or width white dot controls. Holding down the Shift key "locks" the width and length together so they stay in the same proportion as you increase or decrease the diameter of the circle. Figure 6.12 shows the resized cylinder object with a length and width of 1.5in. (meaning that the circle has a diameter of 1.5in.).

You need another cylinder object for the other side of the dog tag, so click the cylinder object you just created and make a copy of it by using the Edit menu or by holding down the proper keys to make a copy (Ctrl+C on Windows or Command+C on Mac). Then select Paste from the Edit menu or press Ctrl+V for Windows or Command+V for Mac to paste a duplicate Cylinder. Drag it to the opposite side of the dog tag, as shown in Figure 6.13.

CHAPTER 6: A Tinkercad Special Project

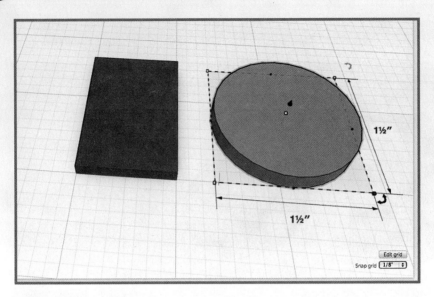

FIGURE 6.12 The cylinder has a diameter of 1.5in.

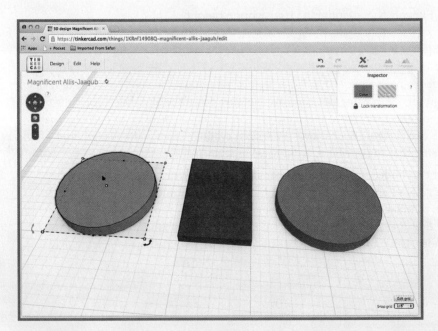

FIGURE 6.13 Create a copy of the cylinder for the opposite side.

Now it's time to place the cylinder objects. Because the Snap Grid setting is set to 1/8in., you'll notice that the Cylinders automatically jump as you drag them around to the nearest

1/8in. grid. Move a single cylinder object so that half of it is overlapping the rectangle, as shown in Figure 6.14.

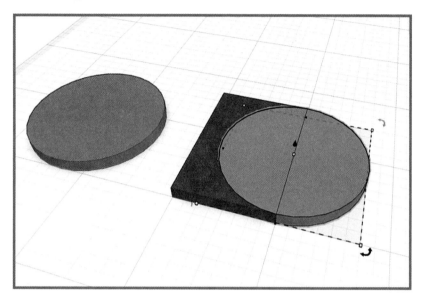

FIGURE 6.14 Place the first cylinder object so it overlaps the rectangle.

Most of the rectangle will be obscured; that's normal. Drag the second cylinder object so that half of it is overlapping the other end of the rectangle. Figure 6.15 shows the final shape of the dog tag, which consists of two cylinder objects and a single box object.

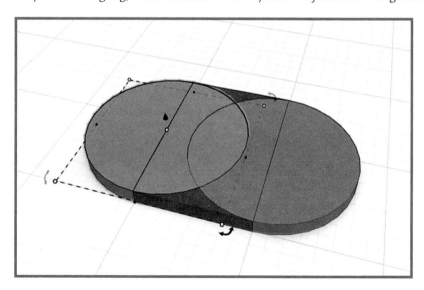

FIGURE 6.15 The second cylinder object, placed over the rectangle.

Rotate the view so you're looking straight down on the dog tag, and you see the overall shape of the final dog tag, as shown in Figure 6.16.

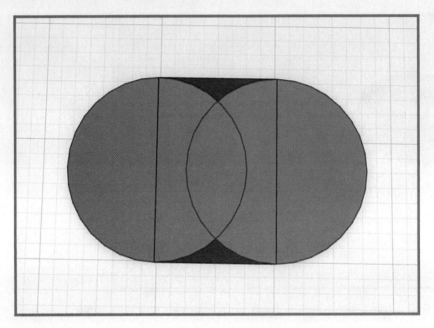

FIGURE 6.16 The basic shape of the dog tag.

At this point, it will be easier to deal with the dog tag as a single object than three objects (two cylinders and one rectangle). You learned about the Group feature in Chapter 5, "Putting Together a Model," and that's what you need to use here. Select all three shapes while holding down the Shift key or by dragging a selection box around the objects and clicking the Group button. All three objects change to a single color (I've chosen yellow) and appear as a solid object (see Figure 6.17).

To finish the basic dog tag shape, you now need to add the small hole at one end. You can do this by dropping in another cylinder object and changing its height to 1/8in., as shown in Figure 6.18.

Creating the Basic Tag Shape

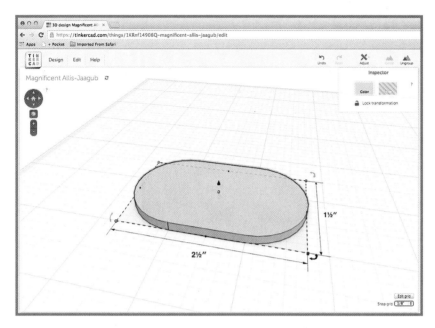

FIGURE 6.17 The dog tag consists of a grouped set of objects.

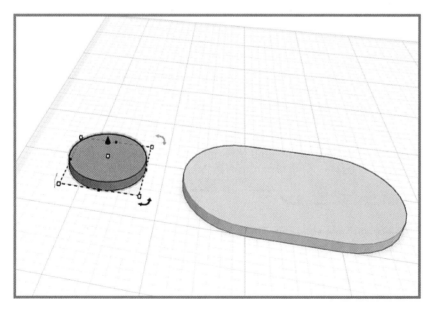

FIGURE 6.18 The final cylinder object will be used to make a hole.

Hold down the Shift key again and resize this new cylinder so it has a diameter of 1/4in. Then click on the Hole button to turn it into a hole object, as shown in Figure 6.19.

CHAPTER 6: A Tinkercad Special Project

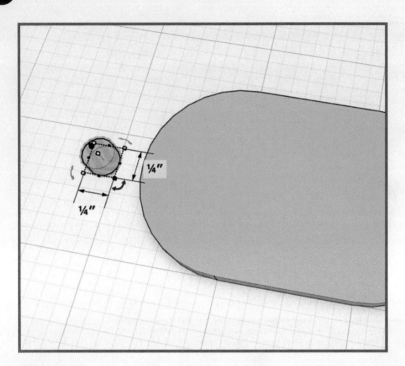

FIGURE 6.19 Resize the cylinder to make a 1/4in. diameter hole.

TIP

You don't have to flatten the cylinder before you make the hole in the dog tag, but by flattening it you'll be able to better see what you're doing as you drag the hole object around for the best placement. A taller cylinder slightly obstructs what's behind it (which might not matter), so flattening to the thickness of the object it will be grouped with allows you to get a better idea of what the final merge will look like.

Drag the new hole object onto the solid tag. The Snap Grid feature helps you properly center the hole in the left edge and with respect to the centerline running parallel to the width of the tag. Figure 6.20 shows the hole object inserted into the tag but not yet grouped.

Creating the Basic Tag Shape

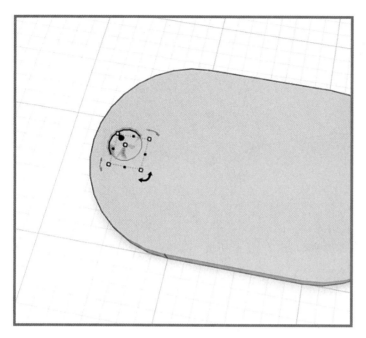

FIGURE 6.20 The hole object is placed but not grouped with the tag.

Now select the hole object and the dog tag object and click the Group button. Figure 6.21 shows that the grouping has been done, and the hole object has cut a clean hole in the dog tag.

Now that the final dog tag shape is complete, it's time to add some embellishments. But before you do that, save this object. When you initially looked at Figure 6.13, you may have noticed that this object has a funny name: Magnificent Allis-Jaagub. Although this is a fun name, it won't really be helpful when you're looking for a specific model. To change that name, click the Design button and select Properties. The Thing Properties dialog appears, as shown in Figure 6.22.

Change the name to Basic Dog Tag Oval - No Text or something similar. This descriptive name will let you distinguish this dog tag model from others, such as a star-shaped or circle dog tag. The "No Text" part of the filename lets you know the tag is blank.

CHAPTER 6: A Tinkercad Special Project

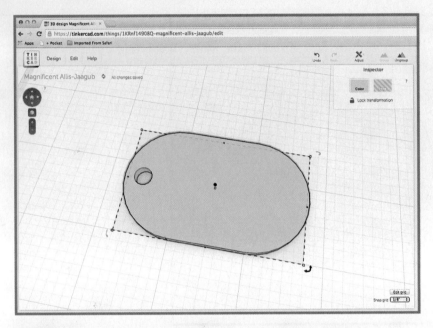

FIGURE 6.21 The finished dog tag shape, complete with hole.

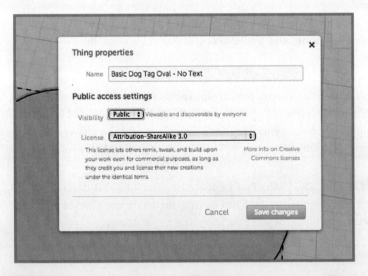

FIGURE 6.22 The Properties window for the new object.

Creating the Basic Tag Shape

TIP

Giving your objects useful names also helps other Tinkercad users who might be searching through the Gallery for something you've created and are sharing with the world. The Tinkercad Gallery is full of models, so someone looking for a basic dog tag shape might have to sort through hundreds and hundreds of models to find your design. Make it easier for them by always giving your objects names that are descriptive and helpful to anyone using the Search tool.

If you change the Visibility setting to Public, other Tinkercad users can use and make changes to your basic shape. If you don't want to share your models with other Tinkercad users, leave this setting as Private. Click the Save Changes button when you're done. Then click the Design button again, followed by Close. You'll see your new model listed on the Dashboard, as shown in Figure 6.23.

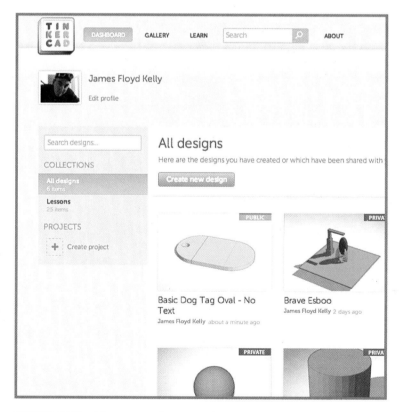

FIGURE 6.23 Your template model is now shown on the Dashboard.

CHAPTER 6: A Tinkercad Special Project

Move your mouse over the thumbnail for the Basic Dog Tag Oval - No Text model and click the small gear that appears. In the drop-down menu that appears (see Figure 6.24), click the Duplicate option.

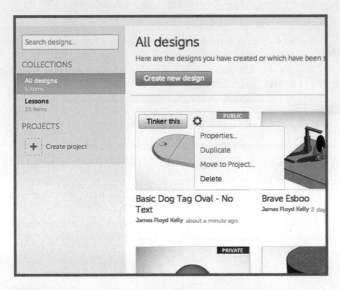

FIGURE 6.24   Duplicate a template object.

When the Copy of Basic Dog Tag Oval - No Text model appears, move your mouse pointer over it and click the Tinker This button (see Figure 6.25). The copy opens to the workspace, and you're ready to start adding some embellishments.

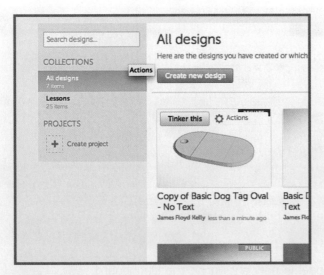

FIGURE 6.25   Open the copy of your dog tag object.

Adding Embellishments

When the Copy of Basic Dog Tag Oval - No Text model opens, click the Design button and select Properties. Change the filename to Dog Tag Oval - Raised Edge and click the Save Changes button.

As you can tell from the new name, you're going to give this object a raised edge. When you're done, you'll save it and then make a copy of it to add even more embellishments.

> **TIP**
>
> Try to develop a habit of saving a copy of a model after you make one or two changes. Doing so will make it easier to go back to an earlier design if you find you don't like a change. In addition, you'll always have a good collection of models in various configurations to choose from when you choose to create a new model.

To create a raised edge on the dog tag, you first need to change the Snap Grid setting from 1/8in. to 1/16in. You'll be removing part of the surface of the dog tag by using a hole object, but if Snap Grid is set to 1/8in., you will be limited to creating a hole object with a minimum height of 1/8in. Then, if you merge that object with the tag by using the Group feature, Tinkercad will create a hole all the way through the tag. By setting Snap Grid to 1/16in., you can create a hole object with a height of only 1/16in. that won't completely put a hole through the dog tag. This will become clear as you work through the next few steps.

To create the raised edge, you need to create another dog tag in the same shape but slightly smaller. How much smaller? In order to get a 1/16in. lip (edge) all the way around the dog tag, the smaller dog tag needs to be 2 3/8in. × 1 3/8in. × 1/16in. (width × length × height). To make this smaller dog tag, create the same three objects as before—a rectangle and two cylinders—with these final dimensions:

- **One rectangle**—1in. × 1 3/8in. × 1/16in. (width × length × height)
- **Two cylinders**—1 3/8in. × 1 3/8in. × 1/16in.

Figure 6.26 shows these three items created and ready to be put together.

Group the objects together and convert them to a hole object by clicking the Hole button. Figure 6.27 shows that the smaller dog tag is now a hole object that is almost ready to be merged with the original dog tag.

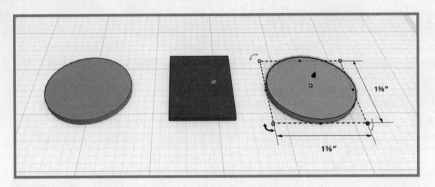

FIGURE 6.26 Objects to create the smaller dog tag.

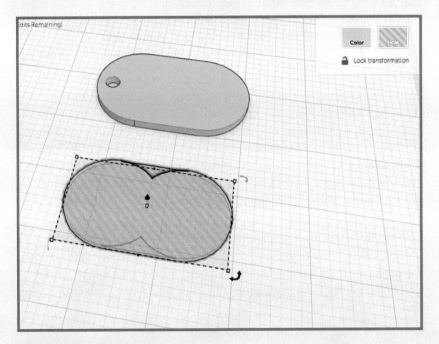

FIGURE 6.27 The smaller dog tag is now a hole object.

The next step is to raise the hole object slightly. Remember that the height of the solid dog tag is 1/8in., and the smaller dog tag (a hole object) is only 1/16in. tall (visible as 0.062" on the screen). You want the raised edge to be on top of the solid dog tag, so you'll need to raise the hole object so its top surface height matches the top surface height of the solid dog tag. Set Snap Grid to 1/32in. so you'll be able to raise the hole object in very small increments and then use the Raise/Lower control to raise the hole object up off the workspace. It will be difficult to tell that the hole object is raised slightly, so you might want to change the view to what's shown in Figure 6.28 in order to more easily see it.

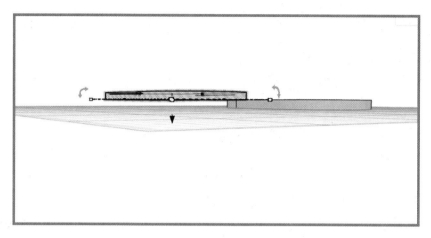

FIGURE 6.28 Rotate the view to more easily see how the hole object floats over the workspace.

Because you're working with decimal amounts, it might be difficult to get an exact match with the surfaces of the solid dog tag and the hole object. Don't worry about an exact match: You just need to sink the hole object into the solid object enough to create that raised edge. You'll be able to change the height of the raised edge later, but for now drag the hole object into the solid dog tag and center it. Figure 6.29 shows the hole object merged with the solid dog tag object.

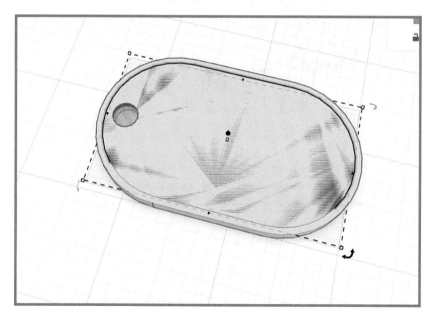

FIGURE 6.29 Merge the hole object and the solid object.

Finally, select both objects and click the Group button. Figure 6.30 shows that the dog tag now has a small lip running along the outer edge.

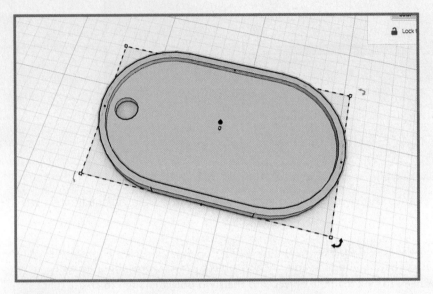

FIGURE 6.30 The dog tag now has a raised edge.

If you're unhappy with the look of the raised edge, simply click the Undo button once to break the grouping and then raise or lower the hole object. Group again and see if you're happy with the result.

Once you're satisfied with the raised edge, save the object and close it to return to the Dashboard. Make a copy of the new project so you won't have to work on the original Raised Edge Dog Tag object. Open that new object, give it a new name, and get ready to finish up this project by adding some text to the dog tag.

Adding Raised Text

Take a look at Figure 6.31, and you'll see that my copy of the dog tag has a nice star-shaped hole in one side. You already have all the knowledge you need to re-create this effect, but I'll share my secrets anyway.

Drop a star object onto the workspace (and decrease its height to 1/8in. if it helps you with placing the hole object properly) while holding down the Shift key to lock in the resizing of the star's length and width so it keeps the star shape. Finally, turn it into a hole object, drag it over the dog tag, and click the Group button.

You could use a variety of the basic geometric shapes in Tinkercad to create your own letter objects, but Tinkercad can save you a little time if you're okay with using its basic

Adding Raised Text

letters, shown in Figure 6.32. (The letter objects are in the toolbar on the left, just below the Geometric section.)

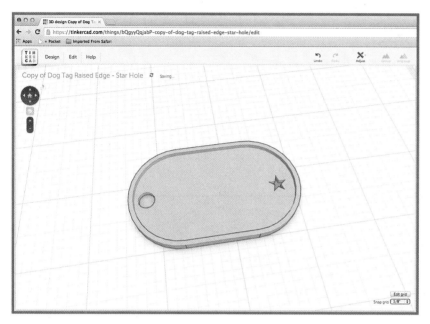

FIGURE 6.31 A new dog tag with a hole and a star, ready for text.

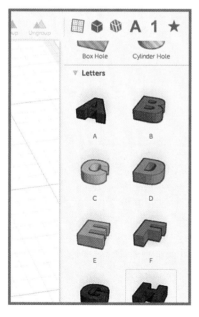

FIGURE 6.32 Letter objects make it easy to add text.

CHAPTER 6: A Tinkercad Special Project

You can resize letter objects just as you can any other objects. You can increase or decrease their height, length, and width, and you can stretch them if you like. For a final embellishment, you're going to add some basic text to the dog tag. You could turn the letters into hole objects so that the negative space they create spells out words, but instead you'll do the opposite here and place the letters so they are slightly raised, like the raised edge that goes around the dog tag.

You can assemble the letters however you like, but you'll need to shrink them in size to get larger words to fit on the dog tag. Set Snap Grid to 1/8in. to make it easier to drag and place the letters nicely. Once you have the word or words spelled out, you can select the letters as a group and shrink them all at once so their spacing and size stay consistent.

Figure 6.33 shows a simple congratulatory message on the dog tag. As you can see here, Tinkercad even offers up some special characters, such as the exclamation point.

FIGURE 6.33   Spell out a short message to place on the dog tag.

Now you just need to raise the lettering so all the letters are 1/8in. in height and then shrink them so they fit properly on the dog tag's surface. Figure 6.34 shows what it looks like after I've shrunk the letters a bit but kept the height of each at 1/8in. In addition, the letters shown here are grouped so they can be easily dragged and dropped onto the dog tag.

Drag your phrase over the dogtag and center it, as shown in Figure 6.35.

Adding Raised Text

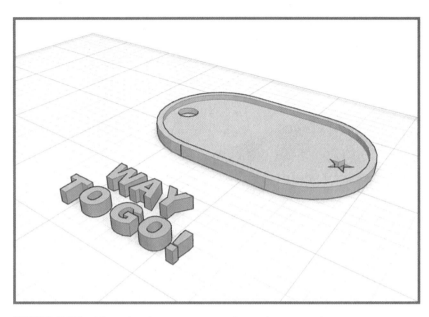

FIGURE 6.34 Size the letters properly and group them.

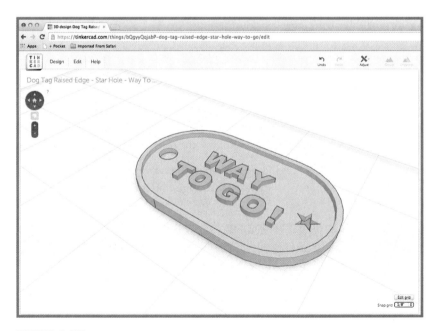

FIGURE 6.35 The letters are now merged with the dog tag.

When you're happy with the location and size of the text, select everything and click the Group button to turn it all into one 3D object, as shown in Figure 6.36.

FIGURE 6.36 One Way to Go! dog tag, ready for printing.

Suggestions for Improvements

There are several ways to turn a 3D model into a real-life object. In Chapter 8, you'll learn how to turn your dog tag object into a real dog tag that you can wear or give to a friend.

Here are some changes or improvements that you could make to this simple little object:

- **Use holes instead of raised letters**—Grouping the letters into an object and then turning that object into a hole object is one way to change things up, and it's fairly easy to do.
- **Create a thin, reversed edge around the dog tag**—By creating two identical dog tag shapes and sandwiching a slightly larger (and very thin) dog tag shape between them, you can create a thin edge to give the dog tag a unique look like the one in Figure 6.37.
- **Create a textured edge**—This embellishment will take a little time, but you can create a textured edge. Start by dropping a small vertical rectangle onto the workspace and turning it into a hole object. Then you can make copies of it and place dozens (or hundreds) of them so they just touch the outside edge. When everything is grouped, the small rectangles create grooves around the dog tag by removing small solid bits of the dog tag.

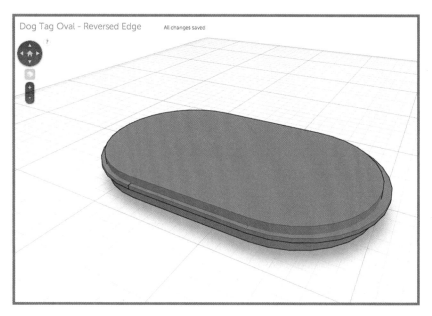

FIGURE 6.37 Another variation of the dog tag with reversed edge.

- **Don't forget numbers**—In addition to providing letters, Tinkercad offers number objects that you can use with your dog tag. With text and numbers, you could create your own ID tag that could contain valuable emergency information, such as a phone number.

Coming up next in Chapter 7, "Another Tinkercad Special Project," you'll learn a few more of Tinkercad's features, create another unique 3D model, and learn some new tricks for creating some eye-catching models.

7

Another Tinkercad Special Project

In This Chapter
- Developing an idea
- Creating a mold for the object
- Creating the elements for the mold
- Finishing up the mold-making project

The secret to getting good with Tinkercad is to keep creating, so that's what you're going to do in this chapter. You're going to learn a few more tricks that will give you even more control over the 3D models you make now and in the future. A lot of professional 3D modelers build up their own libraries of designs that they use over and over again. You started doing that in Chapter 6, "A Tinkercad Special Project," if you created a blank template of the dog tag without the raised edge and another blank template of the dog tag with a raised edge. If you decide you want to create a new dog tag, all you need to do is decide if you want a raised edge or not and then pick the 3D model template that matches your need.

Professional 3D modelers are very fond of the phrase "no need to reinvent the wheel"; they like to be able to save time in their 3D modeling work by tweaking an existing 3D model to their current needs. They end up saving time because they don't have to create a new model entirely from scratch.

Many Tinkercad users have shared their 3D models and made them public (rather than private) so that others can copy their models and then use or modify them. So the community of Tinkercad users helps each other avoid reinventing the wheel.

In this chapter, you're going to learn how to import an existing 3D model into your own Tinkercad Dashboard so that you can use it to create a mold of that model. With the mold, you'll be able to pour in some melted wax to create a solid wax object that will look like the digital model. Along the way, you'll learn a few new Tinkercad tools and tricks that will hopefully inspire you to start creating your own fun 3D models.

CHAPTER 7: Another Tinkercad Special Project

Developing an Idea

Before you can use or modify an existing 3D model, you need to find one. You start by logging into Tinkercad and clicking the Gallery button, shown in Figure 7.1. (If you're still in the design environment, click the Tinkercad logo to return to the dashboard and then click the Gallery button.)

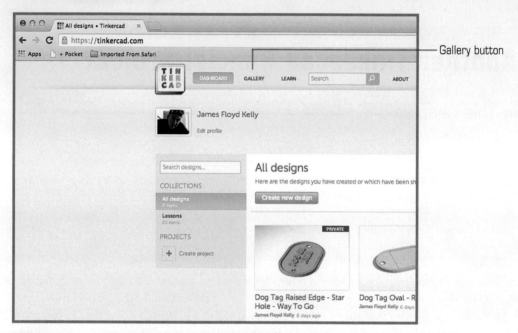

FIGURE 7.1 The Gallery button on the Dashboard.

Clicking the Gallery button takes you to the growing collection of Tinkercad user-created 3D models. There are thousands of them, but unfortunately you can view only 21 per page. Notice that there are four categories that you can use to help narrow down your search for useful models (see Figure 7.2): Hot Now, Newest Things, Staff Favorites, and #template. (There's a fifth category to the far right called My Things, which shows models you have created in Tinkercad. You can also access those models from your Dashboard.)

The Hot Now category shows the most popular models that other Tinkercad users are copying and using. Newest Things shows the models that have most recently been added to the Gallery. Staff Favorites is where the Tinkercad team of developers share their favorite 3D models made with Tinkercad.

Developing an Idea

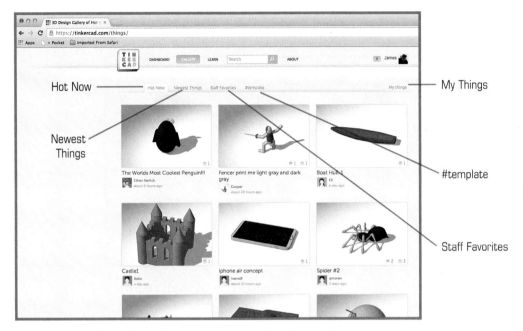

FIGURE 7.2 The Gallery has four categories for models.

> **NOTE**
>
> The Hot Now button also takes into consideration the "Like" feature, so if other Tinkercad users click on the Like button when viewing your models, you may find one of your models appearing on the Hot Now screen!

The #template category shows some great 3D models that Tinkercad users have hashtagged (using #template) and shared. The #template tag makes it easier for Tinkercad users to find models that have been specifically created to be used as templates for other users to copy and modify. You can make copies of models from the other categories, but the ones in the #template category are often much more generic in nature and are meant to be copied and then modified rather than be used exactly as they currently exist.

Many #template models are meant to be used as placeholders for real-life objects. Their creators have often made certain that the measurements of these models exactly match those of their real-world counterparts.

CHAPTER 7: Another Tinkercad Special Project

Regardless of the category you are viewing, scroll down the page a bit, and you see the Show More button (see Figure 7.3).

FIGURE 7.3 The Show More button lets you view more 3D models.

For the project in this chapter, you need a simple 3D model with no parts or pieces that could easily break off if you made a real version of the object. Feel free to find and use any model you like for the remainder of this chapter; when you find an object you'd like to use, click it once, and you see a Preview screen, like the one in Figure 7.4.

FIGURE 7.4 The Preview screen for a Gallery model.

Developing an Idea

You click the Copy & Tinker button if you want to make a copy of a model and add it to your Dashboard. (Don't click it yet!)

I like the model of the chess piece shown in Figure 7.4, but I'm also curious to see what else this Tinkercad user has created. To quickly see what other models a person has created, you simply click that user's name.

After you click the user's name, you see a screen like the one in Figure 7.5, showing the various models this user has created.

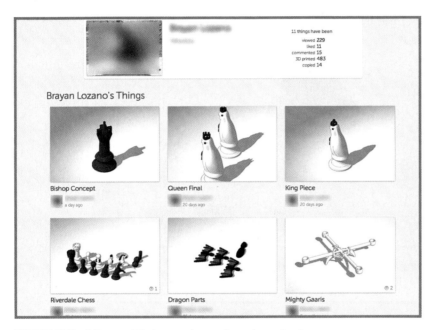

FIGURE 7.5 View a Tinkercad user's other designs.

Let's return to the chess piece model's Preview screen and click the View 3D button to make certain the piece is solid and not hollow. As you can see in Figure 7.6, the bottom shows that the piece is not hollow.

Click the Copy & Tinker button, and Tinkercad opens the model and places it on the workplane, as shown in Figure 7.7. Notice that the object now has a new name: Copy of Bishop Concept.

CHAPTER 7: Another Tinkercad Special Project

FIGURE 7.6 The chess piece is a solid model.

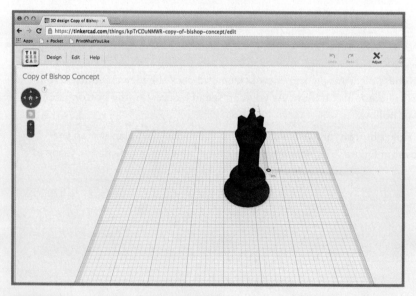

FIGURE 7.7 A copy of a Gallery model, ready for modification.

Developing an Idea

You're going to work on this model throughout the rest of this chapter. To get started with it, you need to change the name of the model and change its orientation on the workspace. So click the Design button and then select Properties. Change the name of the model to Bishop Mold Piece - Solid, as shown in Figure 7.8, and click the Save Changes button.

FIGURE 7.8 Rename the model.

The name of the model is now Bishop Mold Piece - Solid. If you look carefully at the piece, you'll see the letter R placed around the top of the chess piece in four places, as shown in Figure 7.9.

I'm not certain what the letter R signifies, but I'd like to remove it. I'll select the entire piece and then click the Ungroup button until I find that I can click the Rs, select them, and remove them, as shown in Figure 7.10.

CHAPTER 7: Another Tinkercad Special Project

FIGURE 7.9 The letter R around the top of the chess piece.

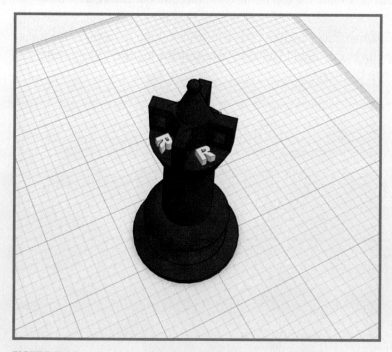

FIGURE 7.10 Remove the Rs from the chess piece.

After the Rs are removed, select the entire chess piece and click the Group button again. You'll be left with the chess piece shown in Figure 7.11.

FIGURE 7.11 Save the modified piece.

You're almost ready to begin working on this chapter's special project, but there's one more step to perform. You'll be modifying this Bishop Mold Piece - Solid object for your new project. Before you make any further changes to it, you should save this object and then make a copy of it (as you did in Chapter 6, with the dog tag templates). This way, you'll have an unmodified standing version of the chess piece that you can return to in the future for additional projects.

To save the project, click the Design button and then click Save. Then close the project by clicking the Design button and then clicking Close. In the Dashboard, move your mouse over the Bishop Mold Piece - Solid project and click the small gear that appears. Then select Duplicate (see Figure 7.12).

After you create a duplicate, open it, rename it Bishop Mold Container, and get ready for your next special project.

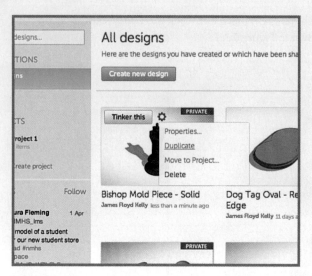

FIGURE 7.12 Make a copy of the standing model.

Creating a Mold for the Object

In Chapter 8, "Printing Your 3D Models," you're going to learn all about how 3D printers can turn a 3D model file (such as the bishop chess piece) into a real object that you can hold in your hands. But there are other ways to turn a digital 3D model into a physical object that you can touch and examine, and one of those is to create a mold of the model. For example, Figure 7.13 shows an example of a Play-Doh mold, like dozens that you probably played with as a kid.

The idea is simple: You open the mold, shove in a small amount of Play-Doh, close the mold tightly and then open it again, and pull out the Play-Doh that now exists in a new form. In this case, the mold in Figure 7.13 is of a superhero. Without any Play-Doh put into the mold, the mold has an empty space inside with the shape of the hero. When Play-Doh is inserted and the mold is closed, the empty space forces the Play-Doh to form the shape of the superhero.

For this special project, you're going to create a mold that, if produced in the real world, would allow you to create a real bishop chess piece from Play-Doh, melted wax, or other substances.

Creating a Mold for the Object

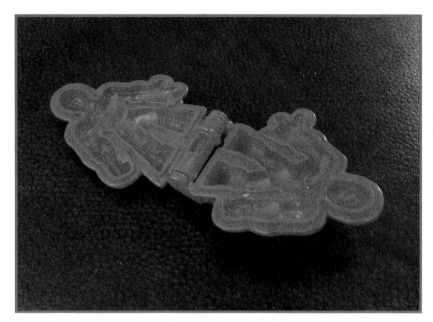

FIGURE 7.13 This Play-Doh mold consists of two halves.

For just a moment, think about what you need to create a mold of the chess piece. Because the chess piece will be the molded object, its shape needs to be the empty space that is surrounded by the solid material that makes up the mold. You already know how to create this empty space in Tinkercad: You simply turn the chess piece into a hole object.

Creating a hole object is fairly simple, but before you do that, you need to add a feature to the object so that you can pour in a material such as melted wax. The material must be liquid in form so that it fills up the empty space before it cools. Once cooled, the mold will be opened, and the object can be pulled out.

One way to pour liquid into a mold is to use something like a funnel. You can easily add a funnel-shaped object to the chess piece object and then group the two objects to create another single, solid object.

To create the funnel, drop a cone onto the workspace, as shown in Figure 7.14.

Select the cone and use the Rotation controls to flip both the cone and the chess piece vertically 180 degrees (along the X or Y axis). Figure 7.15 shows the flipped cone object.

CHAPTER 7: Another Tinkercad Special Project

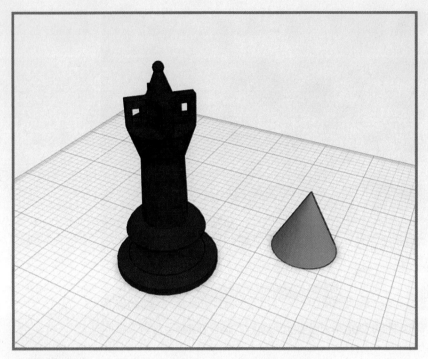

FIGURE 7.14 Use a cone object to create a funnel.

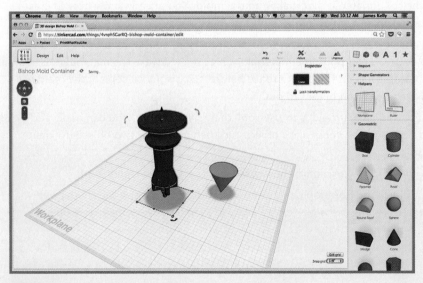

FIGURE 7.15 Flip the cone object to create a funnel.

Creating a Mold for the Object

NOTE

You'll understand shortly why the chess piece was also flipped. The tip of the chess piece is too small to merge with the funnel that will be used to make the mold. The bottom of the chess piece, however, is flat and makes for a great location for the pour funnel.

The next step is to shrink the cone object just a bit. Don't make it too small, or it won't help you pour a liquid into the mold. It's a good idea to shrink it so the diameter of the open funnel is slightly smaller than the diameter of the object. Figure 7.16 shows the cone shrunk just enough so that its widest opening is smaller than the base of the chess piece object.

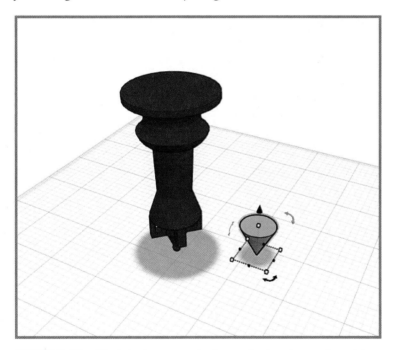

FIGURE 7.16 Shrink the cone before placing it on the object.

You typically pour a liquid into a mold from the top, so you'll now move the cone to the top of the chess piece object. You'll want the cone object embedded in the top so that the opening into the empty shape of the object is wide enough to allow liquid to flow in. Figure 7.17 shows the two objects in different colors, so you can get a rough idea of the size of the hole where the funnel will merge with the chess piece object.

CHAPTER 7: Another Tinkercad Special Project

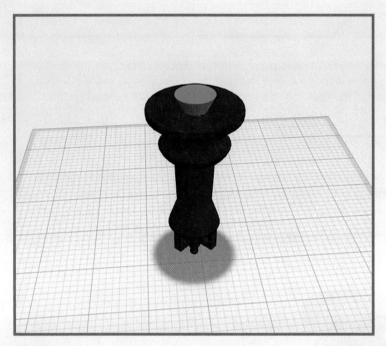

FIGURE 7.17   Merge the cone and chess piece objects near the top.

Use the Raise/Lower control on the cone to raise or lower it into the base of the chess piece. To more easily center the funnel on the base, you can rotate the view of the workspace so that you're looking straight down, as shown in Figure 7.18.

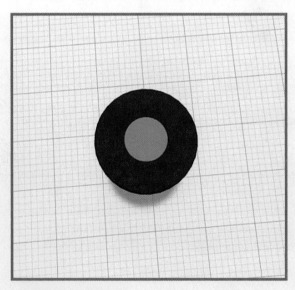

FIGURE 7.18   Center the funnel over the top of the object.

When you're happy with the placement of the funnel, select both objects and click the Group button to merge them into one object. Figure 7.19 shows that the funnel and chess piece are now a single object with the same color.

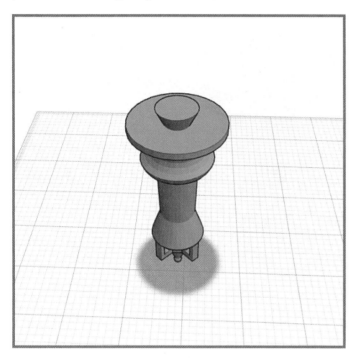

FIGURE 7.19 The chess piece and funnel are now a single object.

Now that the object's final shape has been created, the next step is to create the various elements that will make up the two mold pieces.

Creating the Elements for the Mold

Think about how this mold needs to work: The two halves of the mold will come together so that the empty space inside creates the shape of the final object to be cast. (The word *cast* here means that a liquid or very soft material is forced into the mold and cooled or hardened to create a solid object that can be removed from the opened mold.) If you don't have a way to open the mold, you won't be able to get your solid object out. The easiest way to open a mold is to split it into two parts that can be pulled apart to reveal the object inside.

To create the two mold halves, you have to get creative. There are a few ways to split an object, but here's my favorite method.

First, rotate the view so you're looking at the object from the side. Then make a copy of the chess piece/funnel object and paste it on the workspace, as shown in Figure 7.20.

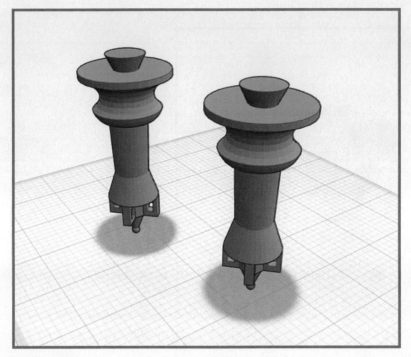

FIGURE 7.20 Rotate the view so you're looking at the object from the side.

Now for a slightly tricky part: Drag one of the objects so its front-left corner matches up to where two of the solid dark blue grid lines cross (see Figure 7.21). It may take a few tries, so be patient.

Next, move the other chess piece/funnel object so that it's also lined up perfectly along one of those dark blue grid line crossings. It should be placed to the right of the other chess piece/funnel object in such a way that the dark blue grid line running from the left of the workspace to the right of the workspace is the common grid line shared by both objects. Figure 7.22 shows that when the two objects are selected, they are lined up along the indicated grid line.

> **TIP**
>
> Put some space between the two objects, but not too much space. I suggest 4 to 5cm (that is, 40 to 50mm).

Creating the Elements for the Mold

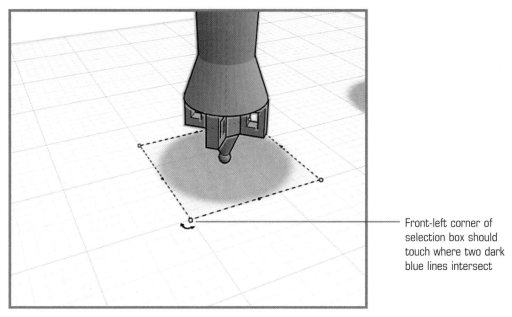

Front-left corner of selection box should touch where two dark blue lines intersect

FIGURE 7.21 Place the front-left corner where two dark blue grid lines cross.

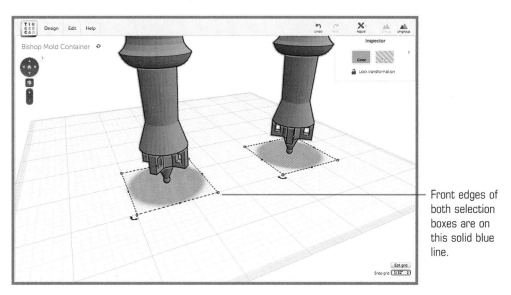

Front edges of both selection boxes are on this solid blue line.

FIGURE 7.22 Line up the objects along a shared grid line.

After you've lined up your objects along a shared grid line, it's time to collect some measurements from the chess piece/funnel objects. As you can see in Figure 7.23, each object is 82.55mm high (not shown in the figure), 38.10mm wide, and 38.10mm long.

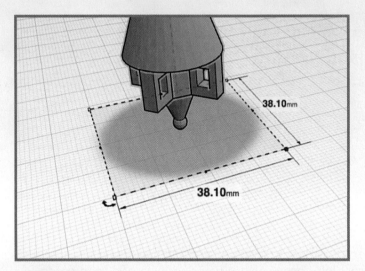

FIGURE 7.23 Get the measurements of the objects.

Now zoom out a bit so you can see the empty space to the left and right of the two objects (while still looking at the objects from the side). You need a box that will surround a chess piece/funnel object but not be too much larger. It must be slightly taller than 82.55mm and slightly longer than 38.10mm, but it needs to be only a little bit wider than half of the object's original width (38.10mm). Drag a box object onto the workspace and resize it so it is 10mm larger than all the other measurements (and all decimal places rounded up):

- **Height**—92mm
- **Length**—49mm
- **Width**—30mm (half of the width is 19.05 + 10mm = 29.05mm)

Figure 7.24 shows such a box object to the left of the chess piece/funnel objects.

Make a copy of this box and drag it to the right of the two chess piece/funnel objects. As you drag the copy away from the original, you see a line extend between the two boxes that will help you align the boxes. You want one box to the left of the leftmost chess piece and one box to the right of the rightmost chess piece, as shown in Figure 7.25.

You now have all the pieces you need to make the mold. You just need to make some slight modifications to the placement of the mold boxes and convert the chess piece/funnel objects to hole objects to create the empty space in the mold halves. Read on to learn how.

Finishing Up the Mold-Making Project

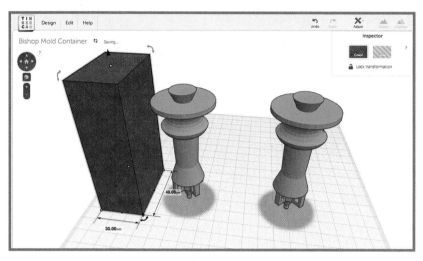

FIGURE 7.24 The mold's outer body is a rectangular box.

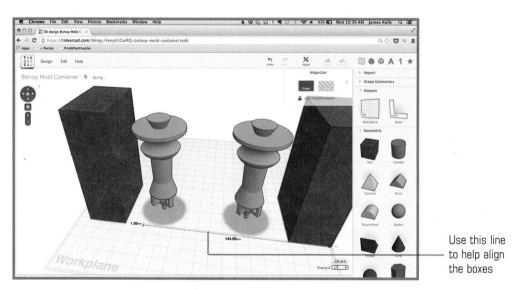

FIGURE 7.25 Place a copy of the box to the right of the chess piece objects.

Finishing Up the Mold-Making Project

Next, you need to slide the boxes back a bit on the workspace so that they "frame" the matching aligned chess piece objects. The easiest way to do this is to keep both boxes selected (by holding down the Shift key) and then slide them forward or back until you're happy with their placement. You might want to rotate the view slightly so you're looking down on the chess piece objects from above, as shown in Figure 7.26. You're going to be centering each chess piece inside a box as best you can. (It doesn't have to be perfect.)

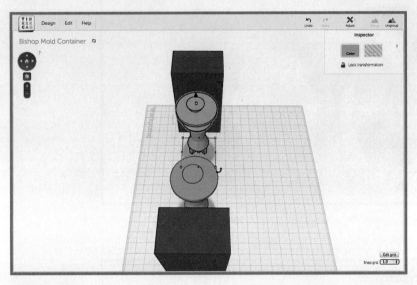

FIGURE 7.26 Rotate the view to line up the boxes.

Now it's time to convert the chess piece/funnel objects into hole objects. Select one chess piece object and click the hole button; then do the same with the other chess piece object. Figure 7.27 shows that the two chess piece objects are now translucent, to indicate that they have been converted into hole objects.

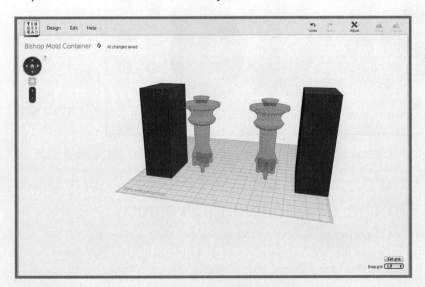

FIGURE 7.27 Convert the chess piece objects into hole objects.

Finishing Up the Mold-Making Project

The next step is to "push" each box toward its respective chess piece object. The box on the left will cover up half of the left chess piece hole object, and the box on the right will cover up half of the right chess piece hole object.

To ensure that your final mold will make a single perfect chess piece, you need to have the left box stop at a point halfway into the left chess piece hole object that will match where the right box stops halfway into the right chess piece.

Now you should understand the importance of aligning each chess piece object along that solid blue grid line. You can use the lighter blue grid lines as a gauge for moving the boxes over the hole objects.

Figure 7.28 shows the left box moved to the right so that it covers half of the left chess piece. Notice that it stops on the indicated solid blue grid line.

Stop on this solid blue line, approximately halfway through object

FIGURE 7.28 The left box covers roughly half of the left chess piece.

Now move the right box to the left, over the right chess piece (holding down the Shift key to help keep the box moving only to the left and right). Keep going until the left edge of the box is resting on the same solid blue line on the right chess piece that bisected the left chess piece in Figure 7.28. Figure 7.29 shows the right-hand box properly placed.

Zoom out a bit, and you should be able to see that each box covers part of the funnel and part of the complete object (see Figure 7.30).

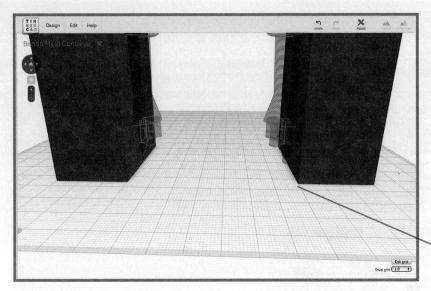

Push the right box to the matching blue line used in previous step

FIGURE 7.29 The right box aligned with the right chess piece.

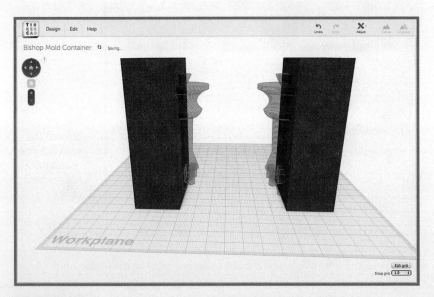

FIGURE 7.30 Each box contains half of the final object.

Recall that your chess piece/funnel objects are 83mm tall, and the boxes are 88mm tall. You're going to need a hole in the top of the mold to pour in the wax or other material, which means you're going to have to lower the boxes by about 10mm so that their top surfaces are parallel and overlapping the top surface of the funnel. To do this, hold down the Shift key and select both boxes. The Raise/Lower control appears halfway between the two boxes, as shown in Figure 7.31.

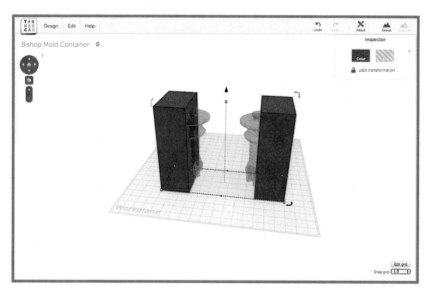

FIGURE 7.31 Use the Raise/Lower control between the two boxes.

Click the Raise/Lower control and drag the boxes down 10mm. Figure 7.32 shows the boxes slightly below the gridded workplane. (This is fine; it does not mean that parts of the boxes have been erased.)

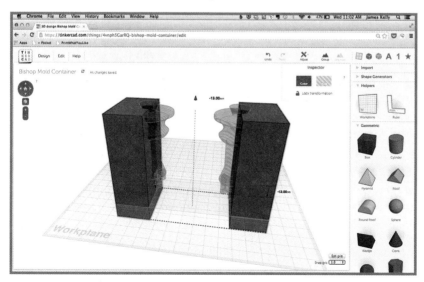

FIGURE 7.32 Lower the two boxes by 10mm.

To finish the molds, you need to drag a selection box around each box and its hole object and then click the Group button. This forces the hole object to remove material from the solid box, leaving behind a mold with an empty space inside. Figure 7.33 shows the final result of grouping together the left box and the left hole object.

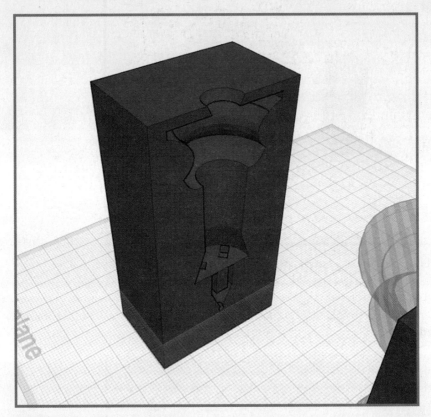

FIGURE 7.33 The left mold, after the box and hole object are grouped.

Perform the same grouping of the right box and right hole object. Figure 7.34 shows the final two mold halves, rotated 90 degrees (and raised 10mm) so you can look inside both at the same time.

The mold still exists as a digital model only, so you can't use it in the real world yet. Figure 7.35 shows how you'd put the two halves of the mold together, with the funnel hole at the top so you can pour melted wax or other material into the mold. Tinkercad even shows you a bit of the inside of the mold when you look down through the funnel hole.

Finishing Up the Mold-Making Project

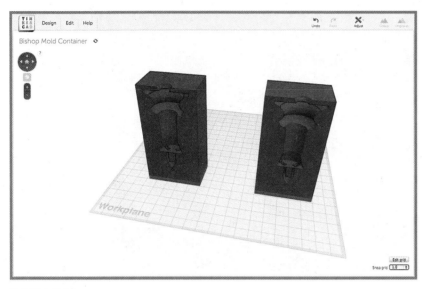

FIGURE 7.34 The final two mold halves.

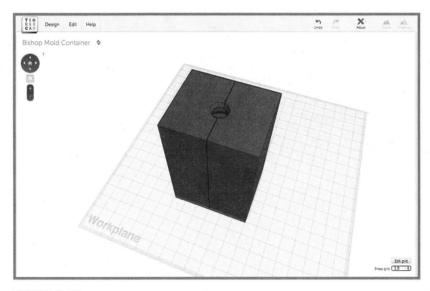

FIGURE 7.35 The funnel is a hole through which you can pour melted wax or other material.

Want to see something really cool? Zoom in on your object, and you'll find yourself diving inside your mold. Figure 7.36 shows what happens when you zoom in just enough to "enter" your mold: You can see the empty space inside in Figure 7.36.

CHAPTER 7: Another Tinkercad Special Project

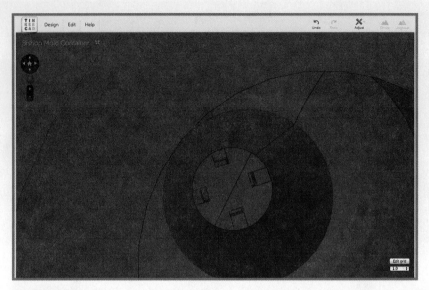

FIGURE 7.36 Zoom in enough, and you can enter the mold.

Congratulations! You've just made a mold of a 3D object. You could use your mold to create real copies of the digital 3D model. I say *could* because right now the mold exists only as a digital 3D model.

To actually make a real object, you need to have a physical copy of the two halves of the mold, right? How do you get those? Chapter 8 shows you how. In that chapter, you'll get a brief break from Tinkercad so you can look at some of the technology out there for turning digital 3D models into real objects you can touch and use.

Printing Your 3D Models

In This Chapter

- What is a 3D printer?
- Creating an STL file
- Melting that plastic
- Moving the nozzle
- Using software to control a 3D printer
- Summary of 3D printing

Up to this point, you've been designing 3D models using Tinkercad and saving them in your Dashboard. There they will stay until you wish to view them, modify them, or delete them. Digital models are fun to create, and they're also fun to show off, especially to friends and family. But in order to share your digital models, you do need a computer that's connected to the Internet; there are times when you don't have a computer or Internet access and so can't log in and share your creations.

Professionals often use 3D modeling to create prototypes of products before they're ever manufactured. They can try out different designs and colors and features before ever creating a version of the prototype they can hold in their hands. But once they reach the point when it's time to create a physical object, what do they do?

A number of tools can turn digital creations on a computer screen into physical objects that can be held, weighed, painted, and even crushed or bent (possibly to test where the item might fail when used by a customer). These tools include Computer Numerical Control (CNC) machines, lathes, laser cutters, vacuum molds, and other advanced (and often expensive) tools. Sometimes, however, pros go the inexpensive route, with someone using a knife to cut the item from wood. Other times, cardboard or poster board is used to create the physical model.

There are many ways to convert a digital model into a physical model. One of the most popular ways is to use a 3D printer. In this chapter, you're going to learn all about this amazing tool that can take a digital object and print it out in plastic and other materials. You end up with a real object you can hold in your hands that matches the dimensions and features of the digital model.

A decade ago, a 3D printer would have run you $50,000, $100,000, or even $1,000,000 or more, depending on the quality, size, and number of items you needed to create. But today, a 3D printer can be had for as little as $300. Even the more advanced versions are no longer out of reach. For $1,000 to $3,000, a school or home can have its very own high-quality 3D printer, capable of connecting to a computer and turning digital into real. Let's take a look at how this happens.

> **NOTE**
>
> I normally avoid self-promotion, but I am quite proud of the beginner-level book I wrote on 3D printing that includes chapters on building an actual 3D printer from a kit, performing a print job, and much more. The book is titled *3D Printing: Build Your Own 3D Printer and Print Your Own 3D Objects*. Visit quepublishing.com for more details.

What Is a 3D Printer?

Think for a moment about an inkjet or laser printer and what it creates. Normally, you put in a blank piece of paper, and out comes that same piece of paper with images, text, or both on it. Whatever the printer puts on the paper, however, is two-dimensional; it's flat. No matter how many times you run that piece of paper back through the printer, new ink is simply placed over the old ink, and you simply end up with a messy-looking piece of paper.

A 3D printer works much like an inkjet or laser printer, but it creates three-dimensional objects instead of putting ink on paper. There are many different types of 3D printers that use a variety of materials to create objects, but they all create real objects from digital models.

This chapter introduces you to a 3D printer that prints using plastic instead of ink. Yes, plastic is typically a hard, solid material, but if you heat it to a specific temperature, it changes to a more liquid state. Have you ever seen someone decorate a birthday cake by squeezing out icing and writing the words "Happy Birthday" on the cake? The decorator moves his or her hands while squeezing the icing out. The end product is a nice handwritten message on a cake, using icing, as shown in Figure 8.1.

What Is a 3D Printer?

FIGURE 8.1 A message on a cake, written in icing.

The cake's flat surface is the workspace, and the decorator's brain sends signals to the muscles in his or her arms and hands to write out "Happy Birthday." The icing is soft, and it comes out the nozzle on the end in a continuous stream, as long as the decorator applies constant pressure.

A 3D printer works much the same way. It has a brain—a small circuit board—that connects to a set of muscles—the motors—that control where the nozzle moves. In this case, however, icing isn't coming out of the nozzle; rather, out comes melted plastic that cools almost instantly when it hits the flat print bed (like the cake's flat surface).

3D printers come in a variety of shapes and sizes, but most plastic-printing 3D printers have a number of items in common. Most of them have a flat print bed, motors, an electronic "brain" of some sort, and a way to feed solid plastic into one end of a device (called a hot end) that heats and melts the plastic before pushing it out another end (the nozzle) and onto the workspace.

CHAPTER 8: Printing Your 3D Models

Figure 8.2 shows one model of 3D printer, called the Printrbot Simple.

What you can't see in Figure 8.2 is that two other motors and the controller are tucked under the base and hidden from view. Take a look at Figure 8.3, and you'll see one of the earliest homemade 3D printers, called the RepRap Darwin, which has just about everything visible.

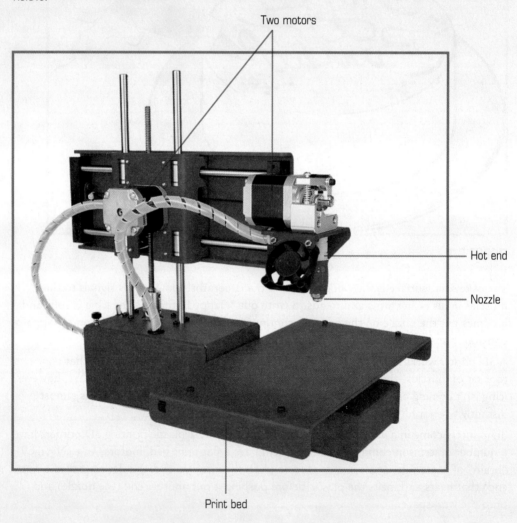

FIGURE 8.2 The Printrbot Simple is a low-cost 3D printer.

What Is a 3D Printer?

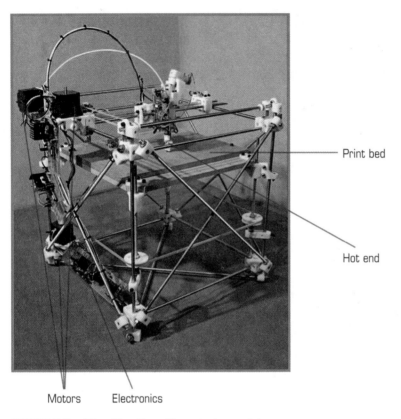

FIGURE 8.3 The RepRap Darwin has all its components exposed.

The RepRap family of 3D printers is a collection of machines that are typically intended to be built by hand. Many hobbyists like the RepRap series because they can build these machines using low-cost parts and upgrade them over time with parts such as more powerful motors. The downside is that building a RepRap printer often requires a certain comfort level working with electronics and extreme patience in debugging and testing the machine to work out the kinks.

In contrast to 3D printers like RepRap are the consumer-friendly models that come fully assembled and ready to use. These out-of-the-box 3D printers are typically a bit more expensive than the build-your-own varieties like RepRap, but they're ready to use quickly: Just take such a 3D printer out of the box, plug it into a computer, and you are ready to go. One popular version of a 3D printer that comes preassembled and ready to go is the MakerBot Replicator 2, shown in Figure 8.4.

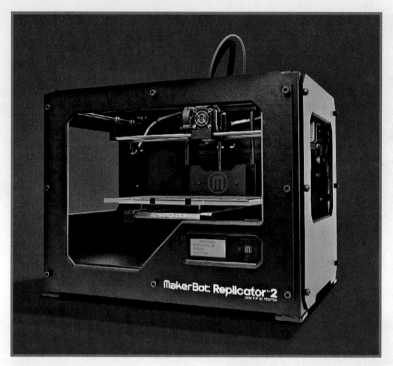

FIGURE 8.4 MakerBot sells an assembled printer called the Replicator 2.

Prebuilt 3D printers like the Replicator 2 tend to be less overwhelming to novices because their cases hide much of the inner workings from view. The Replicator 2, for example, has a simple door on the front that allows you to reach in and pull out a printed object. Most of the electronics and motors are tucked away and hidden from view.

Whether you're looking at a build-it-yourself 3D printer or an assembled one, they all tend to have a few things in common:

- They use a special digital file that is created from the 3D model.
- Plastic is fed into an opening, melted, and then extruded onto a flat workspace.
- Motors are used to move the hot end/nozzle around as the melted plastic comes out in a continuous flow.
- Special software is needed to use the 3D model file properly and instruct the motors where to move as the nozzle extrudes.

We'll now look at each of these in order and examine how they relate to the function of a plastic-printing 3D printer.

Creating an STL File

Open up any of your digital models in Tinkercad and click the Design button. Among the other options listed, you should see the Download for 3D Printing option, shown in Figure 8.5.

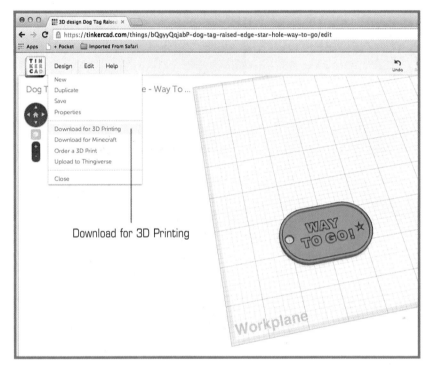

FIGURE 8.5 The Download for 3D Printing option.

Click this option, and a new window appears, offering a number of options for saving your creation to your computer (Figure 8.6). You're interested in the STL button right now.

> **NOTE**
>
> In Chapter 11, "Expanding Tinkercad's Usefulness," you'll learn about 3D printing services. You can send a file of your 3D object to one of these services, and it will print it for you—or cut it using a laser cutter. Some of these companies use different file types, and the extra buttons in the Download for 3D Printing dialog allow you to save a file in some of those other file types.

CHAPTER 8: Printing Your 3D Models

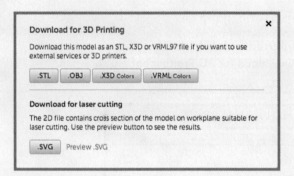

FIGURE 8.6 Click the STL button to get a 3D printer-friendly file.

You're probably familiar with certain file types such as .DOC or .JPG—these are called file extensions and they're how your computer's operating system knows which application to launch when you double-click a file. An STL file is simply a 3D model file with the .STL file extension. Certain applications (such as Repetier, which you'll learn about shortly) are able to open STL files and use them for a variety of things. Objects that you design in Tinkercad will most often be saved as STL files if you want to perform other actions with your model, such as printing it or importing it into more advanced applications.

Click the STL button, and your 3D model is downloaded to your computer. Figure 8.7 shows that this file is named after the 3D model, with the .STL file extension.

FIGURE 8.7 The STL file is stored on your computer.

One additional point you should be aware of when using Tinkercad is the orientation of your digital object. Take a look at Figure 8.8 and notice the word *Workplane* in the lower-left corner. The placement of this word is useful because it tells you what Tinkercad considers to be the "front" of an object.

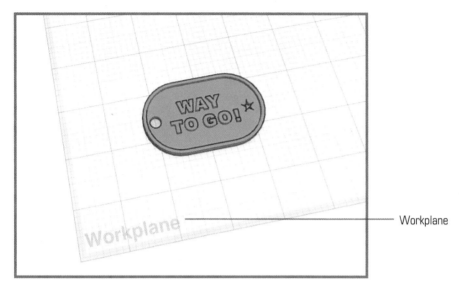

FIGURE 8.8 Tinkercad treats this location as the front of the workspace.

With the rocket object you created in Chapter 5, "Putting Together a Model," it really doesn't matter from which direction you look at the object. Except for the fins, it pretty much looks the same from the front, rear, left, or right. But think about printing an object like the owl object shown in Figure 8.9.

Notice that the word *Workplane* appears behind the owl in Figure 8.9. If this object is saved as an STL file, a 3D printer will be able to print the object, but as it prints, the back (or rear) of the owl will be seen from the front of the 3D printer.

Many 3D printers let you observe your object being printed from any side, but some printers (such as the Replicator 2) have a big window in the front you can look through to watch your object being printed, and the case blocks your view of the printing from the side or rear. Some objects are symmetrical or mostly symmetrical (such as the rocket), while others are not. If you're printing an object and are concerned about one specific side of it (such as the front of the owl object), you'll want to modify the object's orientation before saving.

If you always rotate your object so its "front" is pointed toward the *Workplane* edge in Tinkercad, when you print the object, you'll be able to observe the side that you're most interested in. Figure 8.10 shows that I've rotated the owl object so its face is now facing toward the front of the workspace.

FIGURE 8.9 Some objects aren't symmetrical and will have a "face"/front side.

FIGURE 8.10 Rotate the model, if necessary, to face forward.

Once you have an STL file of your project, you can send it to a plastic-printing 3D printer to create a 3D object. (Most other CAD applications have a way to save a file as STL, but you may need to dig around or read the user guide to figure out how.)

Later in this chapter, you'll see how this file is used to turn your digital model into a physical model, but first let's take a look at how a 3D printer melts solid plastic and then extrudes it onto the workspace.

Melting That Plastic

Figure 8.11 shows a thick loop of plastic and a larger amount on a spool (like thread). This is called *filament*. When you order plastic for a 3D printer, this is what you get.

FIGURE 8.11 Plastic filament comes in small loops or on spools.

Some 3D printers have a small holder where you can place the spool so it spins freely, and other 3D printers require you to come up with a method to feed the thread of thin plastic into the hot end. Either way, a 3D printer typically has the filament going into the 3D printer from the top, as shown in Figure 8.12.

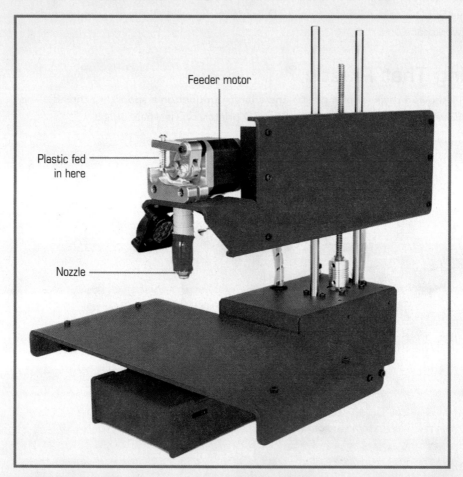

FIGURE 8.12 Plastic goes in up top and comes out below.

The feeder motor consists of a mechanism that grabs tightly on the filament and pushes it down into the body of the extruder. The extruder connects to the controller of the 3D printer and receives electric current to heat up a small cavity inside the extruder. As more plastic is forced into the extruder, the melted (liquid) plastic is forced out the nozzle. In this way, a continuous bead of melted plastic freely flows out the end of the nozzle.

The software used to control the 3D printer won't let the print job start until the extruder reports to the controller that a specific temperature has been reached. This temperature is different for different types of plastic, but once the appropriate temperature is reached, the plastic inside begins to melt, and the feeder motor starts forcing additional plastic, which starts the extrusion process.

Figure 8.13 shows a close-up of the bead of plastic coming out the nozzle. Notice that the bead is much thinner than the original filament being fed in.

FIGURE 8.13 A thin bead of plastic coming out the nozzle.

If the 3D printer's nozzle remained in one position, you'd end up with a real mess. The bead of plastic would exit the nozzle and begin to build up on itself in an ugly pile. To create physical objects, the printer must move that nozzle around such that the extruded plastic bead goes down on the workspace in an orderly fashion—and that's where the other motors of a 3D printer come into play.

Moving the Nozzle

Think again about writing "Happy Birthday" on a cake with icing and how the cake decorator's hands move the icing exiting the nozzle. It's similar to writing your name in cursive on a piece of paper with a pen, but the decorator is simply using icing instead of ink.

When it comes to 3D printers, you can think of the nozzle as the end of a pen. Writing your name in cursive requires that the end of the pen move on the paper without lifting (well, except for dotting *i*'s and crossing *t*'s, but we'll get to that in a moment). When you're writing your name in cursive, your brain instructs the muscles in your arm and hand to move in such a way that you end up with a legible name instead of a scribbled mess.

But how does a 3D printer "write" on the workspace? Remember back to our earlier discussion of the three axes: X, Y, and Z. A 3D printer has a motor for each axis; some 3D printers even have multiple motors for certain axes. These motors move the extruder/nozzle in three different directions. Figure 8.14 shows a typical 3D printer, viewed from the front. From this vantage point, the nozzle can move in six directions: left and right, forward and backward, and up and down. These directions correspond to the three axes: The X motor (hidden underneath the print bed) moves the nozzle left and right; the Y motor moves the nozzle forward and backward; and the Z motor moves the nozzle up and down.

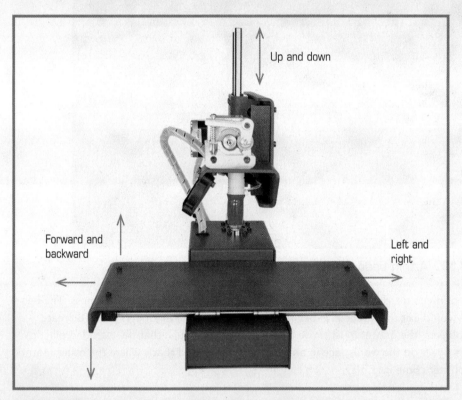

FIGURE 8.14 The nozzle can move in six directions.

Different 3D printers use different methods for actually moving the extruder/nozzle, including threaded rods, string, and belts. Figure 8.15 shows the Printrbot Simple from the side. Notice that the Y axis motor is visible here.

The Y motor has an axle that spins and a small attachment on that axle that has teeth. These teeth fit into the notches on that belt, so that when the Y motor axis spins, the belt is pulled or pushed, forcing the nozzle to move forward or backward.

Moving the Nozzle

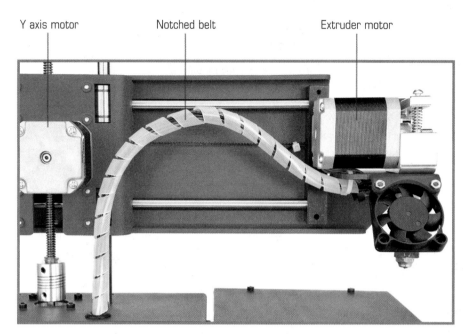

FIGURE 8.15 The Y motor moves the nozzle forward and backward.

The same method is used for the X axis: Beneath the print bed is another notched belt that moves when the X axis motor spins forward or backward.

Now think about how just these two motors can be used to force the nozzle to write your name on the print bed in melted plastic. If the X and Y motors rotate forward and backward faster and slower, they will move the nozzle left and right and forward and backward to trace a specific path.

Until you see a 3D printer in action, this might not make complete sense, so if you have time, point a web browser to the following video to see how the motors work together to move the nozzle around the workspace: http://vimeo.com/73670166.

But what about the Z motor? What does it do? Remember that a 3D object has not just width and length but also height. The Z motor rotates and moves the nozzle up and down, allowing the nozzle to place a bead of melted plastic on top of a cooled bead of plastic that is already on the workspace. This is called layering, and is how 3D printers print objects in all three dimensions.

The Z motor must move the nozzle up (away from the workspace) in smaller increments because the diameter of the bead of plastic being extruded is tiny. For the nozzle to move up in small increments, most 3D printers use a threaded rod to move the nozzle. The threaded rod, shown in Figure 8.16, allows for much finer control in the Z axis than a notched belt.

CHAPTER 8: Printing Your 3D Models

> **NOTE**
>
> Not all 3D Printers have a Z-axis motor to raise and lower the nozzle. Some 3D printers raise and lower the actual printbed. Be aware of this if you begin shopping for a 3D printer and do some research on forums and 3D printing websites to see what customers are saying about a Z-axis motor or a printbed that raises and lowers.

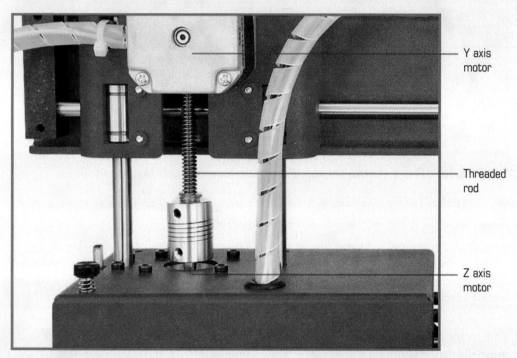

FIGURE 8.16 The Z axis motor turns the threaded rod for up/down movement.

As the X, Y, and Z motors each turn, they determine where the tip of the nozzle is at any given moment. Remember that as these motors are moving, a continuous bead of plastic is exiting the nozzle. If you want to create a simple cube or rectangle to hold in your hands, that bead of plastic will first lay down a single layer of the cube. When that first layer is finished, a second layer is put down, followed by a third, a fourth, and so on. After a hundred or more layers (depending on the size of the cube you are printing), you end up with a solid cube or rectangle, as shown in Figure 8.17.

FIGURE 8.17 A printed rectangular object.

As you can see in Figure 8.17, the individual layers are stacked on top of one another, giving the object a semi-rough surface on its edges. More advanced 3D printers reduce this rough appearance by reducing the diameter of the bead, creating very smooth surfaces.

How does a 3D printer know how and where to put down these layers that, when stacked, create a physical object? It's all done with special software, discussed next.

Using Software to Control a 3D Printer

This is a book about Tinkercad, so I can't go too deeply into the various types of software that are used to print physical objects. Fortunately, the special software that 3D printers use all tends to work the same.

I'm oversimplifying this a bit, but printing a 3D model basically involves just a few steps:

1. Heat the extruder to the proper temperature.
2. Read the STL file and slice the 3D model up into layers.
3. Calculate the best path for the nozzle to move as it extrudes plastic.
4. Communicate the path to the motors so they can follow it.

CHAPTER 8: Printing Your 3D Models

With some 3D printing software, all these tasks are performed within one application. With other software, you need to split up the tasks using two or more applications. For example, some users will have the STL file sliced into layers and saved as a file that the motor-controlling software can read.

Again, I don't have the space to cover every possible option, so instead I'm going to show you one example of software that performs all these options. It's called Repetier, and you can see the basic look of the open application in Figure 8.18. (You can download and experiment with the 100% free Repetier application by visiting www.repetier.com—if you find it useful, consider donating to the software developer to encourage further improvements and features.)

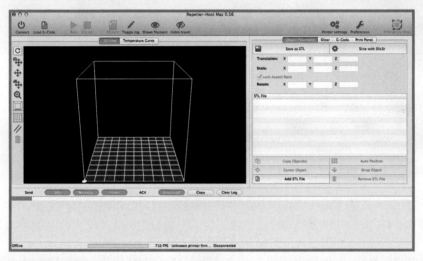

FIGURE 8.18   The Repetier application, open and ready for an STL file.

The gridded box on the left in Figure 8.18 represents the print bed of a 3D printer. Currently there is no object to be printed, but you can easily change that. If you open the STL file you saved earlier, the object appears in the center of the Repetier grid box, as shown in Figure 8.19.

The digital object you created and saved as an STL file is now visible in this box. Notice that it is placed in the box and oriented to match how it was saved in Tinkercad.

TIP

Remember that you should rotate an object if it has a "front" (such as a face) so that it is facing forward. Doing so makes it easier to see the printing details from the front of the 3D printer, especially if you are looking through a small window on the front of the printer.

Using Software to Control a 3D Printer

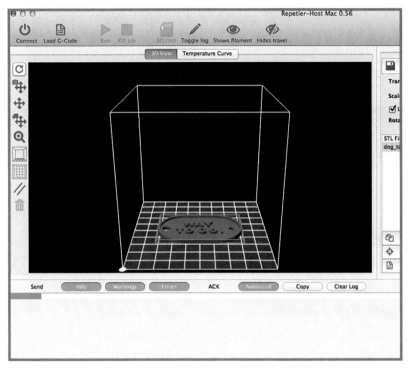

FIGURE 8.19 An STL file places the object to be printed in the grid box.

The Repetier software is also capable of displaying the temperature of the hot end, as shown in Figure 8.20. You use the Repetier control panel to set temperatures and much more.

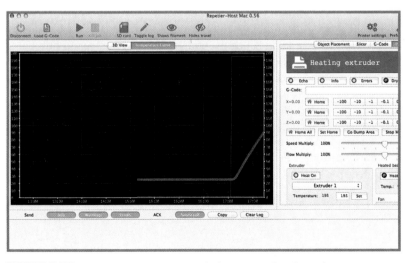

FIGURE 8.20 The temperature of the extruder is going up.

CHAPTER 8: Printing Your 3D Models

While the temperature is increasing, you need to slice the STL file into layers. Figure 8.21 shows that the object has now been converted into a single blue line that represents the path the nozzle will follow.

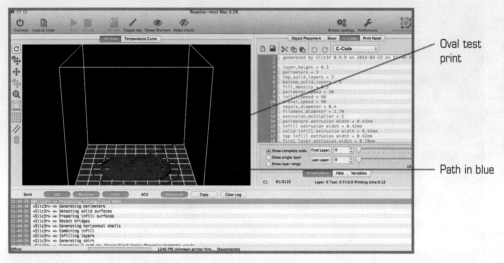

FIGURE 8.21 The object has been sliced and a path calculated.

Surrounding the sliced object in Figure 8.21, notice the slightly larger oval. This oval allows you to observe that the plastic bead is actually being put down and gets the extrusion process going. If you don't see that bead going down, you should cancel the operation and figure out what's going on before trying again.

> **NOTE**
>
> Fine-tuning a 3D printer is a never-ending process, and you'll find an unlimited number of online discussions from hobbyists sharing their tips and advice on proper nozzle placement, distance from printbed, software tweaks, and so on. If you own a 3D printer (or purchase one), consider consulting that company's online forum or tech support for assistance in fine-tuning your 3D printer. It really is different for each model, but rest assured it's not all that difficult to do. I typically spend at least a few minutes for each print project testing and making software settings based on advice from other owners and my own experience.

Notice also in the right side of the window in Figure 8.21 a numbered list of strange words and code. This is G-code, special instructions that are provided to the extruder and motors to set the temperature, the starting point, the ending point, and everything in between. The G-code can consist of a few dozen lines or thousands, depending on the complexity of the object you are printing.

After the software slices the model into layers, you click the Start or Print button to get the printing process started. The temperature starts to rise, and once it reaches the proper melting point for the plastic, the nozzle begins to move. At the same time, the extruder motor begins to rotate slowly, forcing more filament in and melted plastic out of the nozzle. The nozzle moves left, right, forward, backward, and up (as additional layers are printed), turning your digital model into a physical object.

If everything goes as planned, the printer finishes the print job, and a small plastic object ends up sitting on the printer for you to pick up.

Summary of 3D Printing

As mentioned earlier, I've simplified the 3D printing process a bit in this chapter. It really isn't all that difficult, but a lot of little things can go wrong. Time and practice help a 3D printer user find the right settings that allow for consistent printing and good quality.

3D printers come in a variety of sizes and shapes, and the prices vary. And many 3D printers don't just print using melted plastic. There are 3D printers that use lasers to melt powder, and there are 3D printers that print in chocolate. There are many variations, and if you're looking to buy one, you can spend hours or days sifting through all the different options. To start comparing different 3D printers and the various methods they use for printing, head to www.3ders.org, shown in Figure 8.22. At this website, you can read articles on the workings of 3D printers as well as view price comparisons of the various 3D printer manufacturers and sellers.

CHAPTER 8: Printing Your 3D Models

FIGURE 8.22 3ders.org offers articles, price comparisons, and more.

In addition to using a 3D printer, there are other options for turning your 3D models into physical models. You'll learn about some of those options in Chapter 11.

In Chapter 9, "More Useful Tricks with Tinkercad," we return to Tinkercad and dive into a few more tricks for creating 3D models.

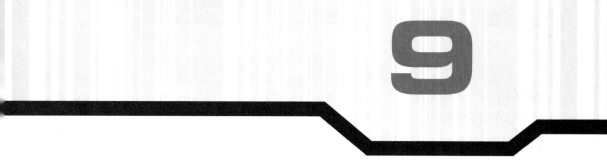

More Useful Tricks with Tinkercad

In This Chapter
- Using the mirror feature
- Importing your own sketch
- Experimenting with the shape generators tool
- Where to next?

Tinkercad is an easy-to-use CAD application. Its simplicity means that the application doesn't overwhelm novices with hundreds of buttons and menu options and toolbars. Some of the most advanced CAD applications have hundreds and hundreds of buttons and menu options to choose from, and that can scare away someone who is new to 3D modeling. As you get started with 3D modeling, you can push your skills as far as possible with the tools that Tinkercad offers; when you discover a limitation, you may decide it's time to start examining more advanced CAD applications.

You've learned quite a bit about Tinkercad so far in this book, but there are still some fun and useful tricks left to learn. In this chapter, you're going to see three additional features Tinkercad offers and how you can use them with your 3D models. Two of them will be immediately useful, and the third will require you to spend some time experimenting before you really get the hang of it.

Up first is one of the easiest options available; it can ultimately save you tons of time in designing 3D models.

Using the Mirror Feature

When you look at a mirror, you see a reflection of yourself. You look like you, but everything is reversed. If your hair is parted on the right, the "you" in the mirror has hair parted on the left. Wink your left eye, and the person looking back winks the right eye. Mount a mirror above you, and all of a sudden the "you" in the mirror is upside down, looking down at you.

CHAPTER 9: More Useful Tricks with Tinkercad

Mirrors can help you work in CAD applications like Tinkercad. The best way to see how this feature works is simply to play around with it a bit until it makes sense. Go ahead and open a new project in Tinkercad and drag the following three objects onto the workspace: a sphere, a box, and a pyramid. You can see these three objects lined up in Figure 9.1.

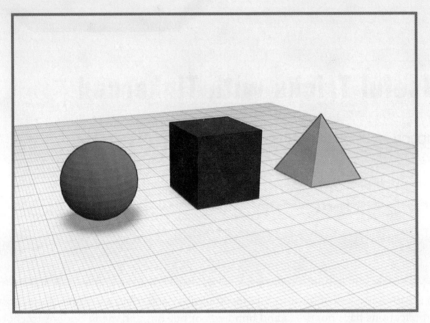

FIGURE 9.1 Three independent 3D objects placed on the workspace.

Use the Raise/Lower tool to place the pyramid on top of the box. Then merge the sphere with the front face of the box. Next, click the Group button to create a single object like the one shown in Figure 9.2. (Your final object doesn't have to look exactly like this one—just similar.)

Next, select the new object, copy it, and place three copies of it on the workspace, as shown in Figure 9.3.

Now, to see how the Mirror feature works, select one of the objects by clicking it and then click the Adjust button. You see the familiar Align option in the drop-down menu, but now you want to select the Mirror option (see Figure 9.4).

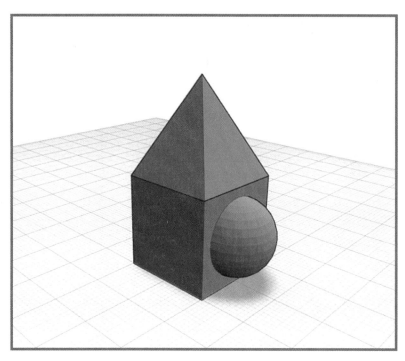

FIGURE 9.2 Create a single unique shape by grouping.

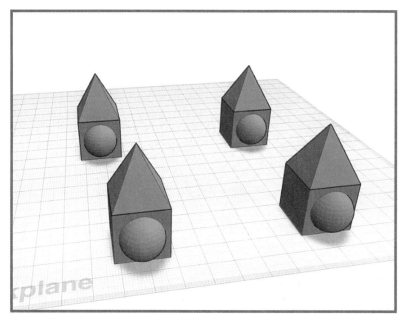

FIGURE 9.3 Create three copies and spread them all around the workspace.

CHAPTER 9: More Useful Tricks with Tinkercad

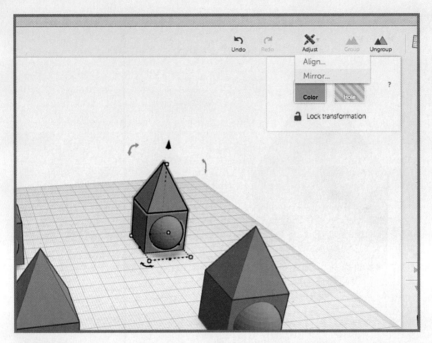

FIGURE 9.4 Select an object and then choose the Mirror option.

After clicking the Mirror option, three solid lines with arrow heads on each end appear around the selected object, as shown in Figure 9.5. These are the Mirror controls.

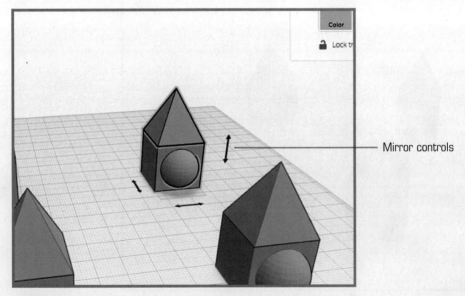

FIGURE 9.5 Mirror controls appear surrounding an object.

You can click one of these Mirror controls to modify the selected object. Notice that each line has two arrow heads that point in opposite directions. If you imagine placing a mirror in the place where an arrow head is pointing, you'll begin to understand how the controls work. In Figure 9.6, for example, clicking the indicated line forces the object to change so that it looks like a reflection seen with a mirror placed behind it.

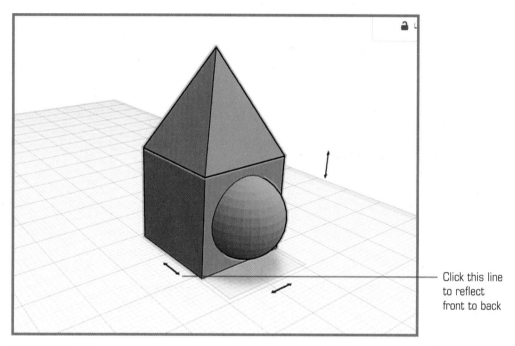

Click this line to reflect front to back

FIGURE 9.6 The arrows will "reflect" the object in the selected direction.

A mirror doesn't actually appear onscreen, but Tinkercad does draw a yellow outline of the object's final orientation for you to view before you actually click the line. Just hover your mouse pointer over an arrow, and you see the yellow outline of the object, as indicated in Figure 9.7.

Go ahead and click the arrow, and the object "flips" so that its final orientation matches what would be seen in a mirror placed behind the object. In this case, the sphere on the front is now on the back of the object; Tinkercad has reversed the object (not rotated it) so that it's a mirror image of the original (see Figure 9.8).

CHAPTER 9: More Useful Tricks with Tinkercad

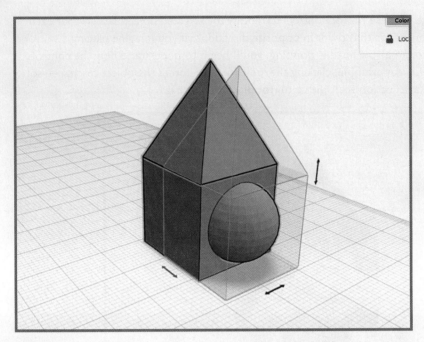

FIGURE 9.7 A preview of the object's mirrored position appears.

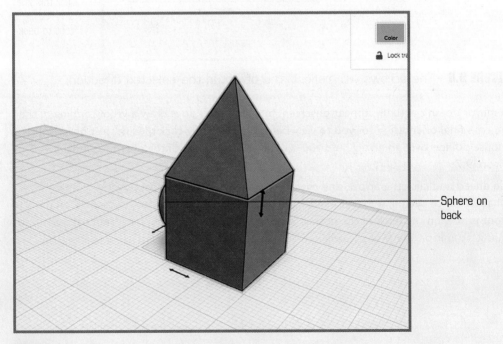

FIGURE 9.8 The object changes so it is a mirror image of the original.

You can click the Undo button once to undo the mirroring action or simply click the same arrow you chose before to mirror it back to its original shape.

In some instances, clicking an arrow has no effect. What does a sphere reflected in a mirror look like? Right: a sphere. Mirroring the object in Figure 9.9 left to right won't change the way it looks; the Pyramid on top and Sphere on front are symmetrical from left to right, so you won't see the object's yellow outline that indicates changes that will be visible.

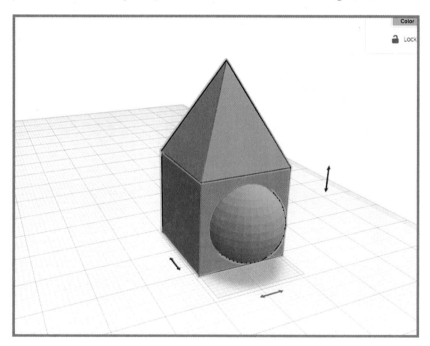

FIGURE 9.9 Sometimes the mirror action has no effect.

You can mirror the object top to bottom by clicking the arrow indicated in Figure 9.10. Then the yellow outline shows how the object will flip to match a shape reflected in a mirror placed above or below the object.

While you can easily use the Rotate controls to turn an object from front to back or completely upside down, the Rotate tool doesn't always give you the effect you're looking for. Take a look at Figure 9.11, which contains half of a smiley face and an exact copy placed beside it. Say that you want to merge these two halves to create a whole face.

CHAPTER 9: More Useful Tricks with Tinkercad

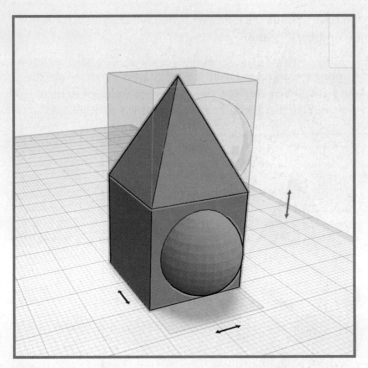

FIGURE 9.10 You can mirror an object top to bottom.

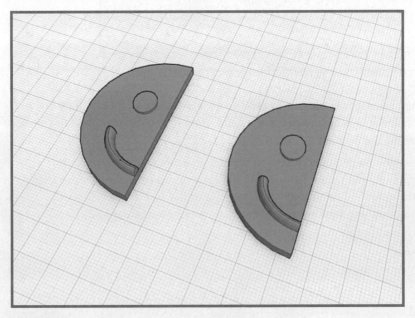

FIGURE 9.11 Two halves that need to be merged.

If you simply select one half and rotate it 180 degrees on the Z axis, you won't end up with the desired effect, as shown in Figure 9.12.

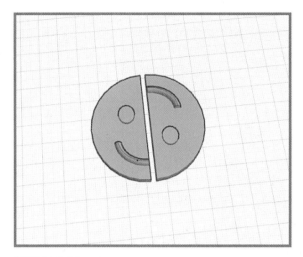

FIGURE 9.12 Rotating around the Z axis doesn't produce a mirror image.

Likewise, rotating on the Y axis simply flips one of the half smiley faces over so you're looking at its underside, as shown in Figure 9.13.

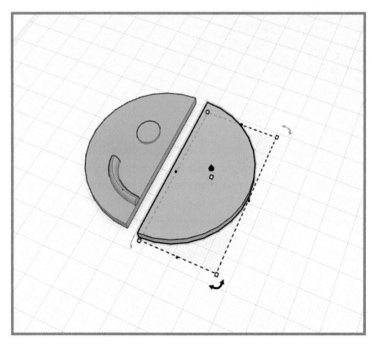

FIGURE 9.13 Rotating on the Y axis also fails to produce the desired effect.

Only using the Mirror controls allows you to take a copy of one half of an object and "reflect" it so that joining the two halves produces a truly symmetrical match, like the one in Figure 9.14.

FIGURE 9.14 The Mirror controls allow you to create symmetrical pairs of objects.

Think about this for a moment: The Mirror controls allow you to take a copy of an object and reflect it so the two halves together create a symmetrical object. This can be a truly time-saving tool!

If you're trying to create a super-complex 3D model that will be symmetrical along the X, Y, or Z axis (or two axes, or even all three), you might consider creating only half (or one-quarter or one-eighth) of the object in Tinkercad and then making a copy of that object and mirroring the copy on the desired axes.

If you choose to create a 3D model of an hourglass, for example, you can simply create the upper or lower half of the hourglass, make a copy, and then mirror it on the Z axis (up and down) and merge the two halves. Modeling a spaceship? Create the left side of it (looking at it from the front), copy it, and then mirror the copy on the Y axis (side to side) before merging.

Using the Mirror controls will not only save you time but can help you ensure that the two halves of your object are identical in every way.

Importing Your Own Sketch

Tinkercad offers a variety of standard shapes (called primitives in CAD speak), such as the cube, pyramid, and ring objects. You drop these on the workspace; adjust their length, width, and/or height; merge them with other objects; group them; change their colors; and use other objects to create holes that are merged and grouped. You end up with 3D models that are created by blending and combining basic shapes with solid objects and hole objects. It works well, and if you put in the time, you can create some truly amazing 3D models.

You may have noticed, however, that Tinkercad lacks a line-drawing tool. Later in the chapter, you'll see an advanced method for creating unique-shaped objects, but if you want to quickly draw something in Tinkercad like the shape shown in Figure 9.15, you're just out of luck.

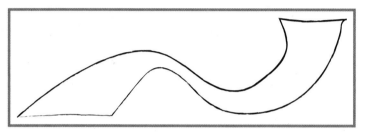

FIGURE 9.15 A simple hand sketch of a unique shape.

You could probably duplicate the look of this unique shape by merging the basic primitives in Tinkercad with hole objects, but the process could be very time-consuming. It would be nice if Tinkercad offered a way to sketch a shape and import it. Fortunately, this is an option, but you need to do a little planning to take advantage of this option.

In Chapter 10, "Where Can You Find Existing Models?" you'll learn how to import existing STL files from websites that share 3D models from designers all over the world. It's similar to the Gallery models available to you in Tinkercad, but these STL files aren't hosted by Tinkercad, so you must save them to your computer before pulling them into Tinkercad. But how do you do that?

Take a look at Figure 9.16, which shows the Import section of the toolbar that runs down the right side of the screen.

The Import section allows you to pull in an STL file or an SVG file from your computer (or a website, if you use the URL option). Tinkercad can open either of these file types and then place the shape or object it contains on the workspace.

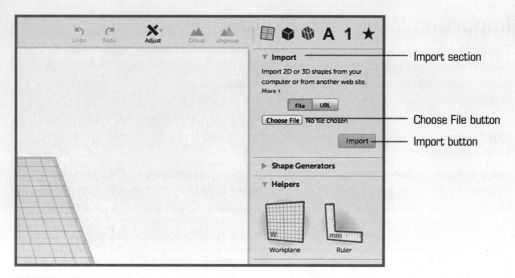

FIGURE 9.16 The Import section allows you to pull in shapes and models.

You'll see how STL file importing works in Chapter 10. Here you'll see how the SVG file option lets you import a basic hand sketch.

SVG stands for *scalable vector graphics*. You may be familiar with other graphics files, like PNG, GIF, TIFF, or JPG. SVG is another standard file format, and it's the only one that Tinkercad accepts besides STL. This means that if you want to import a hand-drawn shape you've created, you need to have it saved as an SVG file. How do you do this?

Some—but not all—graphics programs enable you to save a drawing as an SVG file. The good news is that if you can save a hand sketch as a JPG, you can find tools online to convert that file to SVG. Let's look at how it's done.

Figure 9.17 shows a very simple shape I've created in a graphics program called Skitch. To make a similar shape yourself, you can use any program that is capable of saving a sketch as a JPG.

After creating the image, I saved it as a JPG on my desktop and opened a web browser to the website www.online-convert.com, shown in Figure 9.18.

Importing Your Own Sketch

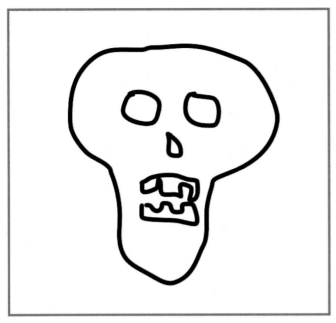

FIGURE 9.17 A hand-sketched shape created in a graphics program.

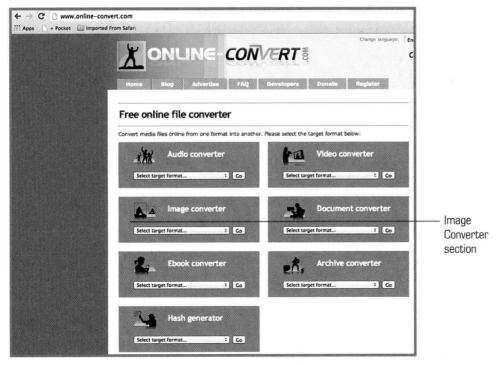

FIGURE 9.18 Use the www.online-convert.com website to change a JPG to SVG.

Click the Image Converter section's drop-down menu and select the Convert to SVG option, as shown in Figure 9.19.

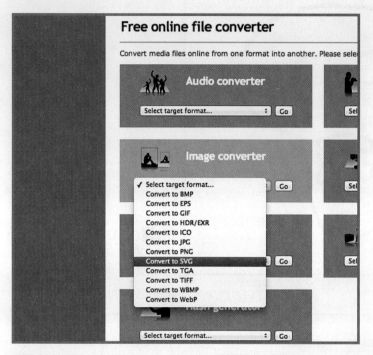

FIGURE 9.19 Choose the Convert to SVG option from the drop-down menu.

A new screen like the one in Figure 9.20 appears. It allows you to browse to the JPG image you saved on your computer.

Click the Choose File button and browse to the location of your JPG image. Select the Monochrome option and then click the Convert File button. The new SVG file is saved to your computer, as shown in Figure 9.21.

Return to Tinkercad and click the Import section's Choose File button and then browse to the location of the SVG file. If you created the sketch with a computer application, you're going to need to reduce the size of the imported shape, so enter a value in the Scale section between 10% and 20%, as shown in Figure 9.22.

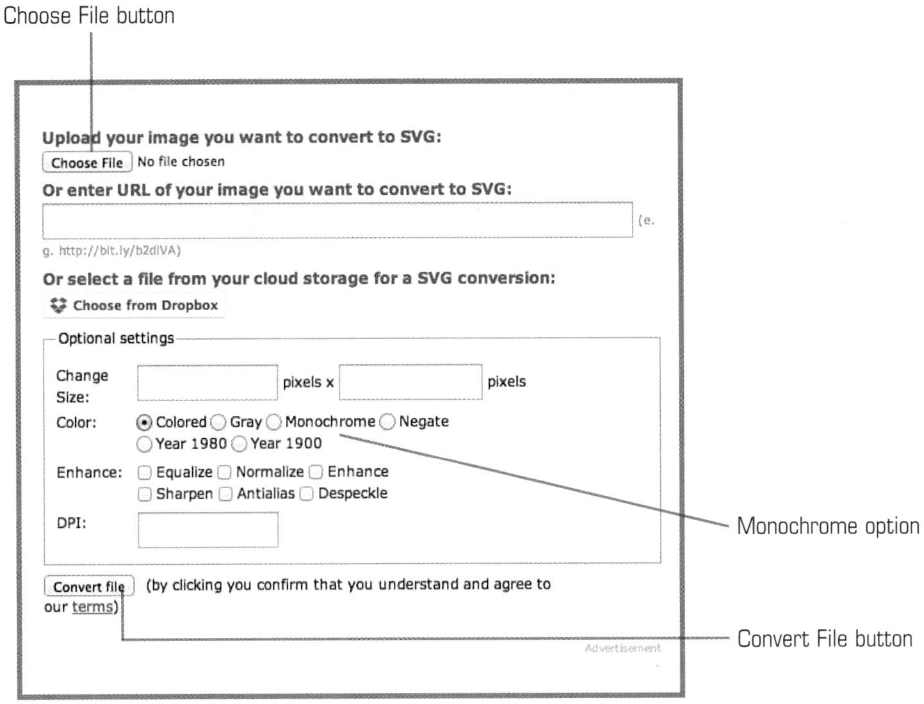

Importing Your Own Sketch

Choose File button

Upload your image you want to convert to SVG:
[Choose File] No file chosen
Or enter URL of your image you want to convert to SVG:
[   ] (e.
g. http://bit.ly/b2dIVA)
Or select a file from your cloud storage for a SVG conversion:
❖ Choose from Dropbox

Optional settings

Change Size: [   ] pixels x [   ] pixels
Color: ⦿ Colored ◯ Gray ◯ Monochrome ◯ Negate
◯ Year 1980 ◯ Year 1900
Enhance: ☐ Equalize ☐ Normalize ☐ Enhance
☐ Sharpen ☐ Antialias ☐ Despeckle
DPI: [   ]

[Convert file] (by clicking you confirm that you understand and agree to our terms)

— Monochrome option

— Convert File button

FIGURE 9.20 Select the file to be converted to SVG.

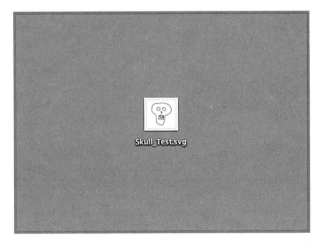

FIGURE 9.21 Your JPG file now exists as an SVG file.

CHAPTER 9: More Useful Tricks with Tinkercad

> **NOTE**
>
> You can experiment with different scale values if you like, but remember that you will always be able to enlarge or shrink your imported object by using the sizing controls.

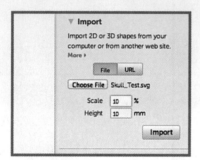

FIGURE 9.22 Change the scale of the imported object.

Click the Import button and prepare to wait anywhere from 30 seconds to a few minutes, as Tinkercad converts the SVG file of your sketch to a digital model. If you've done everything correctly, you should eventually end up with a nice 3D model of your sketch. Figure 9.23 shows my hand sketch converted to a digital model.

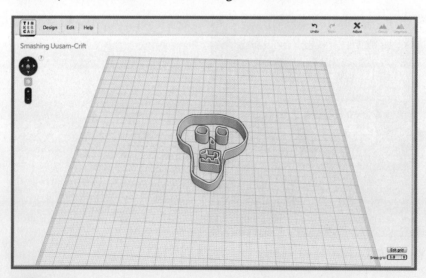

FIGURE 9.23 Your hand sketch becomes a 3D model.

Now you can save your new model, change its color, merge it with other shapes, make it a hole object to cut out its shape from another solid object, or use the Design menu to save it as an STL file that you can print on a 3D printer.

As you can see, it's actually pretty easy to convert a drawing you've done in a graphics program into a file that can be imported into Tinkercad. Keep in mind that if you want the imported sketch to be solid (not hollow, like the skull in Figure 9.23), you need to fill in the drawing with a solid color. If you select the Monochrome option at www.online-convert.com, any solid color you use will be converted to black, and the imported shape will appear solid, as shown in Figure 9.24.

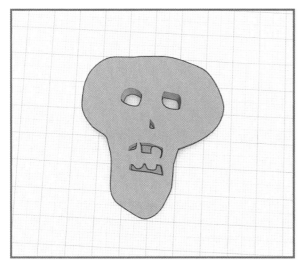

FIGURE 9.24 The filled-in SVG file and the final 3D model.

CHAPTER 9: More Useful Tricks with Tinkercad

Finally, if you have a hand-drawn pen or pencil sketch that you want to import, you need to find a way to scan that sketch and save it as a JPG image. You'll get the best results if the sketch exists on a clean, white background because www.online-convert.com treats anything white as empty space.

Experimenting with the Shape Generators Tool

Tinkercad has a useful tool called the Shape Generators, and you find it on the toolbar running down the right side of the screen, as shown in Figure 9.25. This tool lets users create custom shapes by using the JavaScript programming language.

FIGURE 9.25 The Shape Generators tool lets you create custom shapes.

Programming languages are the tools that software developers (also often called *programmers*) use to create apps for computers, tablets, and mobile phones. Programmers enter special text (and a lot of it!) to create games, websites, and much more. Programming languages are sometimes very easy to learn, and sometimes they're quite difficult. The text

Experimenting with the Shape Generators Tool

shown in Figure 9.26 is an example of JavaScript that Tinkercad uses to create custom shapes. Notice that while it does appear to be English, it's a bit cryptic in many places.

FIGURE 9.26 The JavaScript language allows Tinkercad to create special shapes.

This particular bit of JavaScript (often referred to as *code*) allows you to design a custom ring. You can modify the diameter, the thickness (height), and a few other characteristics by dragging the dots and the drag bars shown in Figure 9.27 to change the shape of the ring.

When you click the Shape Generators tool, it opens to show you some community-shared shapes, as well as shapes you've created, listed in the Your Shape Generators section.

To create a custom shape, expand the Your Shape Generators section as shown in Figure 9.28. Click the New Shape Generator button.

A drop-down menu appears, offering a bunch of preconfigured shapes, including ring, torus, and star. Click one of them, such as your ring, and the JavaScript window appears on the screen.

At this point in your Tinkercad career, all you can do is modify the basic shape by using the drag bars and control dots. But if you know JavaScript, you can make modifications to the shape—for example, adding an extra point or putting a hole in the center of the star.

Why would you want to learn JavaScript for use with Tinkercad? Tinkercad offers a good selection of primitives (cube, sphere, etc.). But what if you wanted to create an eight-pointed star? Think about the work you'd need to do to drag primitive shapes and modify them; think about all the angle rotation changes you'd need to make and the careful placement required to put all those points together to make a nice-looking star. With JavaScript, you could create that same star by using mathematical equations (of a sort) to define where each point is with respect to the others. You could define the thickness of the star and even the angles between each arm. This might sound like just as much work as using Tinkercad to draw the eight-pointed star, but it's really not...if you know JavaScript.

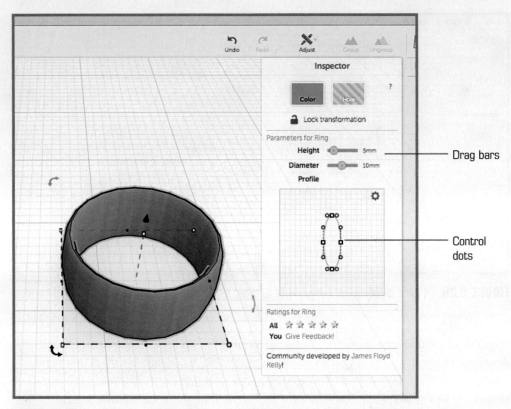

FIGURE 9.27 Use drag bars and control dots to modify a custom shape.

FIGURE 9.28 Click the New Shape Generator button to create your own shape.

> **NOTE**
>
> JavaScript instruction is beyond the reach of this book. If you want to learn more, see one of the many JavaScript books out there.

Where to Next?

Congratulations! As you've worked through the book to this point, you've gotten your hand on almost all the tools and features Tinkercad offers. In Chapter 10 you'll learn about a few more tools in the Design menu. (It is possible that Autodesk might add some new features after I've finished writing this book, so consult the forums or Help option to see what you can find.)

But you're not done yet. You've still got a few more chapters that will give you some more hands-on practice. But you don't have to wait until you read them to get creating. Feel free to take a well-deserved break before starting Chapter 10 to create something new and unique that you can share with the Tinkercad community. Go have some fun!

Where Can You Find Existing Models?

In This Chapter
- Welcome to Thingiverse
- Additional 3D model sources

You've already learned how easy it is to use the built-in Tinkercad Gallery to find 3D models created by other Tinkercad users and copy them to your own Dashboard. The Gallery allows you to easily see what other Tinkercad users are creating as well as to grab copies of the models so you can use them and even modify them to fit your own needs.

But the Gallery isn't the only 3D model repository out there. There are hundreds of websites dedicated to holding 3D models, and the great news is that almost all of those 3D models are available to you as free downloads.

This chapter introduces the granddaddy of all 3D model repositories, Thingiverse, and some of the various fun things you can do from that site. Finally, at the end of the chapter, I provide some additional 3D model repositories for you to investigate on your own.

Welcome to Thingiverse

One of the most popular sites for finding and downloading 3D models is Thingiverse.com. Owned by MakerBot, the company that offers the Replicator 2 3D printer you read about in Chapter 8, "Printing Your 3D Models," this site is currently the largest repository of digital models available for free download.

Figure 10.1 shows the homepage for Thingiverse (www.thingiverse.com). As you can see, the site has a professional look and feel, with a changing banner at the top that informs visitors of new additions and features.

CHAPTER 10: Where Can You Find Existing Models?

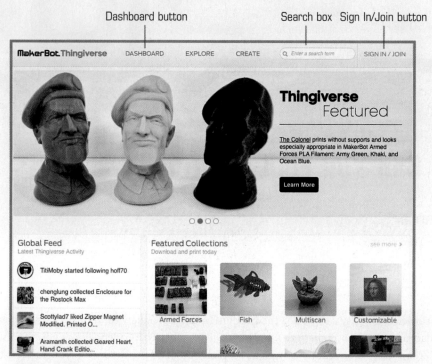

FIGURE 10.1   Thingiverse.com's homepage, with a banner featuring information and new additions.

While browsing and downloading from the Thingiverse library is free and requires no user account, you'll find that creating a free account is helpful for tracking the models in which you're interested. And you must have a user account if you want to upload any of your own 3D models to share with other Thingiverse users.

Click the Sign In/Join button in the upper-right corner and follow the instructions to create an account. When you're signed up, the Dashboard is a useful place to visit to manage your downloads and uploads.

Whether you've created an account or not, you can use the Search box for finding 3D models related to keywords you type in. Looking for a bust or likeness of Abraham Lincoln? Type in "Abraham Lincoln," and you'll get a list of results like the one in Figure 10.2.

When you find a model you'd like to download or examine more closely, simply click it, and an information screen like the one in Figure 10.3 opens.

Welcome to Thingiverse   201

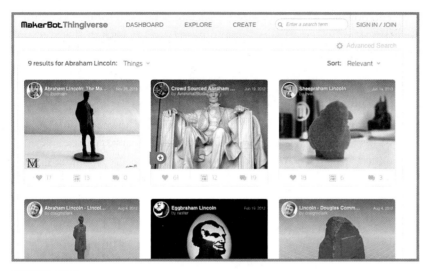

FIGURE 10.2 Use the Search box to find 3D models that interest you.

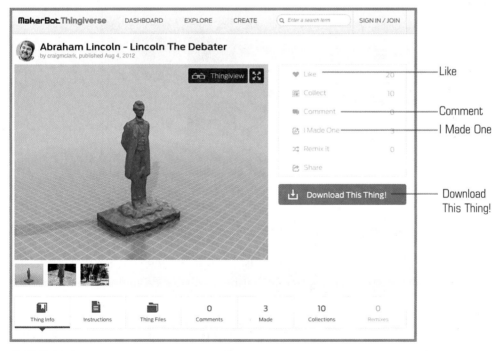

FIGURE 10.3 Open a 3D model's information screen to get more details.

CHAPTER 10: Where Can You Find Existing Models?

From this screen, you can download the STL file (refer back to Chapter 8 for details on the STL file format) by clicking the Download This Thing! button. You can click the Like button to let the object's creator (and others) know you're a fan, click the Comment button to add a comment (if you have a user account), and click the I Made One button to alert the world that you printed out this object with a 3D printer or 3D printing service. (You'll learn about 3D printing services in Chapter 11, "Expanding Tinkercad's Usefulness.")

Scroll down the page a bit, and you see some additional buttons, as shown in Figure 10.4. It can be helpful to click Instructions because some 3D model creators provide special instructions for using a model, such as advice on getting a good 3D print. To the right, in the Makes section, you see user-submitted photos that show the final results of 3D print jobs.

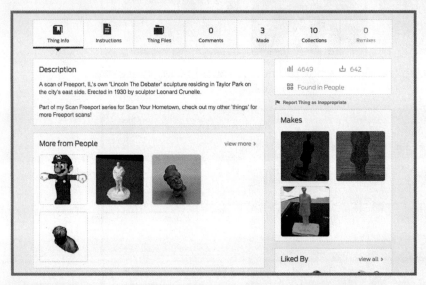

FIGURE 10.4 More options for a 3D model on Thingiverse.

One important thing you need to know about any 3D models you download is what permissions the object's owner has enabled for using the model. Scroll down just a bit more on an object's page, and you see the License box, as shown in Figure 10.5.

A set of small icons give you the basic idea of what you can and cannot do with the model. The crossed-out dollar sign, for example, tells you that you are not allowed to sell the model's files or a 3D printed version of the model. These icons, however, aren't always obvious, so I recommend that you always click the link that describes the licensing. In the case of this Abraham Lincoln model, if I click the Creative Commons – Attribution – Non-Commercial link, I'm taken to the page in Figure 10.6, which explains it in much more detail.

Welcome to Thingiverse 203

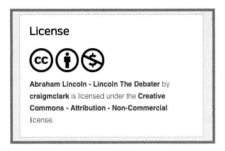

FIGURE 10.5 Check an object's license limitations before proceeding.

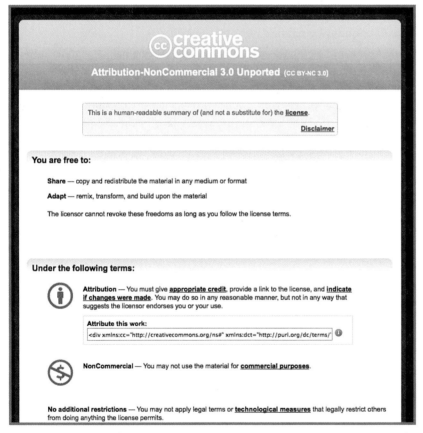

FIGURE 10.6 Click the licensing information link to get more details.

The page in Figure 10.6 tells you that you may download and share the Abraham Lincoln file as well as make modifications to it; the Share and Adapt descriptions provide these details. Don't skip the Under the Following Terms section because this is where you find the rules you must adhere to in order to share or adapt the model. In this case, you must always give credit to the model's creator, provide a link to the original model, and note any modifications you've made to the files. Finally, the NonCommercial section lets you know in very specific terms that you may not profit from the model in any manner.

In a nutshell, with this Abraham Lincoln model, I can and cannot do the following:

- I cannot download this file and then upload it on another 3D model repository and claim to have created this model.
- I can download the Abraham Lincoln model and modify it in Tinkercad as much as I want, including putting a Viking helmet on his head if I like. Basically, I can have as much fun as I like with the file.
- I cannot print out little Abraham Lincoln statues and sell them to my neighbor, to a stranger at the mall, or on my company's website.
- When I upload my Abraham Lincoln model, complete with Viking helmet, I must note in the model's description that the model was created by user craigmclark on Thingiverse and include a link to the original model. (Thingiverse kindly provides this link for easy copying, as you can see in the Attribute This Work box in Figure 10.6.) I must also make note that my model comes with the new and improved Viking helmet addition that (shockingly) the original lacks.

If you can agree to the terms provided by the original model creator, download the STL file to your computer and then import it into Tinkercad. (Importing an STL works the same way as importing an SVG file; refer back to Chapter 9, "More Useful Tricks with Tinkercad," for the steps to perform an import action). Click the Download This Thing button on the model's information page, and the webpage jumps down to the Download All Files button, shown in Figure 10.7.

FIGURE 10.7 Click the Download All Files button to get the 3D model file.

In some instances, 3D model creators may offer variations of their models. You can click the Download All Files button to grab them all, or you can click a single file in the list to download just that file. After you click the button or the file, the download begins.

After you've downloaded the model's STL file (or files), open Tinkercad and use the Import feature you learned about in Chapter 9. Figure 10.8 shows the Abraham Lincoln model open in Tinkercad and ready for tinkering.

FIGURE 10.8 Import the STL file to view the model in Tinkercad.

Make any changes you want to make to the model, including adding the currently popular Viking helmet. When you're done, click the Design menu and save your modified object. Figure 10.9 shows my new and improved model.

> **NOTE**
>
> Remember: If you choose to share this model with other Tinkercad users in the Gallery, you need to give credit to the original model's creator and include a link to the original!

Before you close the model and return to the Dashboard, you need to save an STL version of the Abraham Lincoln model to your computer so you can upload it to Thingiverse. Simply click the Design menu and choose the Download for 3D Printing option and the STL file type. After the file is saved to your computer, you can return to Thingiverse to upload your modified model.

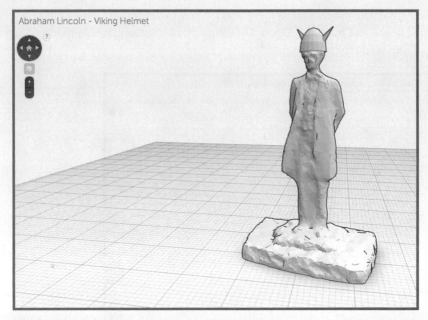

FIGURE 10.9 Modify the model in Tinkercad.

Uploading a model to Thingiverse is fairly straightforward. Log in, if needed, and click the Create button near the top of the screen. You then see a drop-down menu like the one in Figure 10.10.

FIGURE 10.10 Thingiverse offers three different Create options.

If you want to upload a model of your own to Thingiverse, click the Upload a Thing! button. You then see a screen like the one in Figure 10.11. (You should definitely click the other two options to investigate them when you have time, but they're beyond the scope of this chapter.)

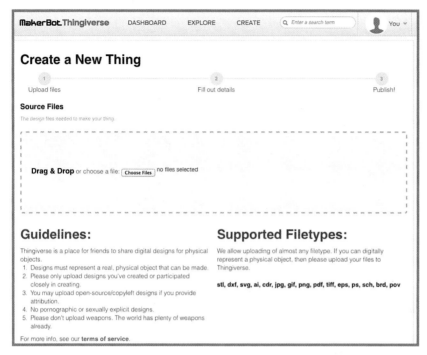

FIGURE 10.11 Step 1 for uploading your 3D model to Thingiverse.

Click the Choose Files button and browse to the location of your model's STL file. The file appears in the list, and you need to scroll down the page a bit to provide the additional information. As you can see in Figure 10.12, I've included the original model user's name and a link to the original model, and I've described the modification I made to the model. I've also chosen the same licensing option—Creative Commons - Attribution - Non Commercial—as for the original. I've also selected a category and included some search words in the Tags section so that any users looking for Viking helmet models will find my new addition.

All that's left is to click the Publish button. If everything works as it should, you see your new model added to the Thingiverse library, complete with its own information page, as shown in Figure 10.13.

Now that you know how to search and download models from Thingiverse as well as upload your own models, you should spend some time looking around at models that interest you. If you click the Explore button at the top of the Thingiverse screen, you can browse the entire library by using keywords, categories, and collections of similar models.

CHAPTER 10: Where Can You Find Existing Models?

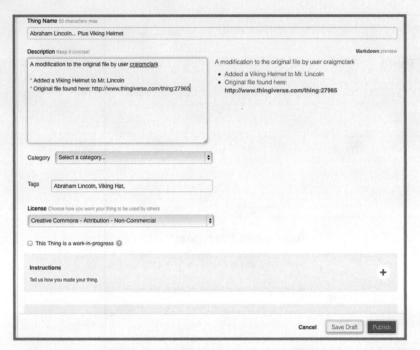

FIGURE 10.12 Fill out the detail information for your model upload.

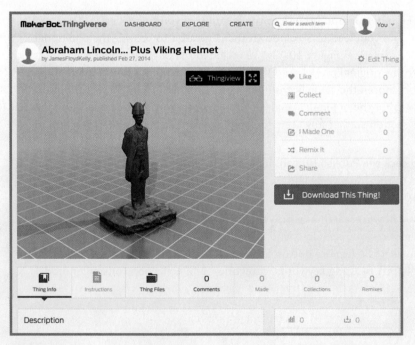

FIGURE 10.13 Your new model is now part of the Thingiverse library.

Additional 3D Model Sources

Thingiverse has plenty to offer you as you continue to expand your 3D modeling skills, including plenty of advanced models for inspiration. But Thingiverse isn't the only 3D model library out there. While Thingiverse has the largest online 3D model library, there are plenty of competitors out there that are growing slowly but surely. Here are just a few 3D modeling libraries for you to investigate—keep in mind that some of them may require you to register for an account:

- fabster.com/physibles
- cubehero.com
- grabcad.com/library
- 3dhacker.com
- repables.com
- youmagine.com

You can find thousands of models to download and use with Tinkercad, and if you have a 3D printer, you can even create physical models from them. But what if you don't own a 3D printer or have access to one? Not to worry: There are plenty of options for turning your digital models into physical models, as you're about to discover in Chapter 11.

11

Expanding Tinkercad's Usefulness

In This Chapter
- Find a 3D printing service
- Taking your object into *Minecraft*

In Chapter 8, "Printing Your 3D Models," you got an overview of how 3D printers work and how they can turn a digital 3D model into a physical model made out of plastic and other materials. Over the past five years, the prices of 3D printers have dropped substantially, and prices will continue to drop as more and more companies enter the 3D printer market. Even better, the 3D printers that are being released for consumers are becoming easier to use so that anyone can take advantage of the ability to print objects when needed.

Even with the drop in prices and improved ease of use, however, not everyone has the budget to purchase a 3D printer. Some people don't want the hassle of dealing with a printer and the filament or other material needed to run it. If you're just interested in creating 3D models for fun, the money spent on a 3D printer might be better used purchasing more advanced CAD software once you've pushed the limits of Tinkercad.

But there's always the off-chance that you might want or need to turn one of your 3D models into a physical model, and when that time comes, you'll be happy to know that you don't need to spend money on a 3D printer. Instead, you can have a 3D printing service print the model for you. You'll save money, the printing service will ensure that the print job turns out well, and the final model will be sent to you all wrapped up in a box.

In this short chapter, you'll learn how easy it is to get a Tinkercad model printed using a 3D printing service. And in case you're a fan of *Minecraft* (a popular game at the moment), you'll also find out how easy it is to convert your 3D models into in-game objects you can import into your *Minecraft* game.

CHAPTER 11: Expanding Tinkercad's Usefulness

Finding a 3D Printing Service

The best way to see how easy it is to use a 3D printing service is to give it a try, so in this section, you'll open up a model and walk through the steps in Tinkercad to order a printed version of the model.

Figure 11.1 shows the dog tag model from Chapter 6, "A Tinkercad Special Project." This model is 2.5in. × 1.5in. × 1/8in., and printing it seems like it would be an easy job for most 3D printers. But let's take a look at what a 3D printing service can do.

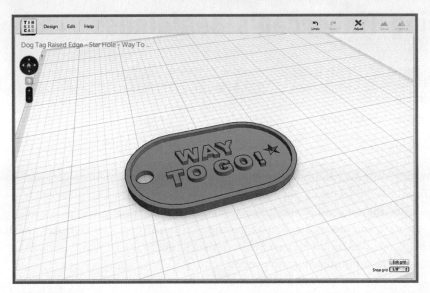

FIGURE 11.1 The final dog tag to be printed as a physical model.

Click the Design menu, and a drop-down menu appears. Click the Order a 3D Print option, as shown in Figure 11.2.

Tinkercad has partnered with several 3D printing services: Ponoko, Sculpteo, Shapeways, and i.materialize. The number of partners may increase in the future, but today, these four are your options—and they're all great options for turning your digital model into a physical model. They all work in a similar way, allowing you to upload your model, pick the material, and select options such as size, color, and a few other choices.

If you click any of the four buttons shown in Figure 11.3, that company's website opens in a new browser window.

Finding a 3D Printing Service

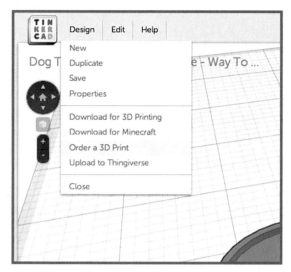

FIGURE 11.2 The Order a 3D Print option is for accessing 3D printing services.

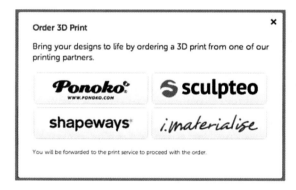

FIGURE 11.3 Four 3D printing partners of Tinkercad.

3D printing services range from rather simple to quite advanced in terms of the various services they offer. For example, if you click the i.materialize button (while your model is displayed in Tinkercad), the model is immediately uploaded to i.materialize, without any login required. Click the Continue Printing button, and you're taken to the i.materialize website, as shown in Figure 11.4.

As you can see, the i.materialize site lets you quickly choose the material and lists the price it will charge. By default, your model is selected at its full size (100%), but you can resize it by using the drag bars if you desire. Smaller is cheaper than larger, and price is also determined by the material and finish you choose for printing. Some of the more expensive material options have even more configuration options.

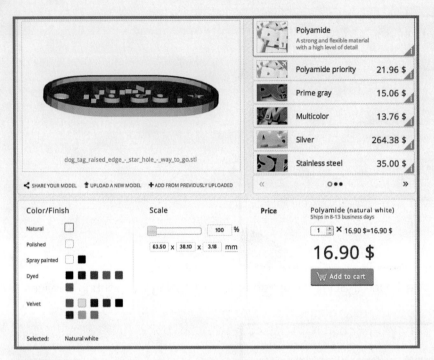

FIGURE 11.4 The i.materialize 3D printing service is simple but easy to use.

Keep in mind that just because the dog tag is green in Tinkercad doesn't mean the final printed object will be green. If you select the first option, Polyamide, for example, the default print job will be done in a white plastic. You can have it printed in green, but adding color will increase the price. Click the Green color or any other color, and you can see the price change. For most printing services, the least expensive option is typically a white or gray color.

If you want to create a lifelike dog tag, you can select the Stainless Steel option (currently $35.00) and choose Gold Plating, for example. You can click the More Options button to see even more options for materials, as shown in Figure 11.5. How about a titanium dog tag for $300? The printing service gives you a printing cost and delivery estimate for each option.

Once you've selected a material and any configuration options, click the Add to Cart button and proceed to the checkout to pay for your order. Figure 11.6 shows what your order might look like at this point, with details on shipping times and costs. If you're satisfied with all the details, click the Proceed to Checkout button and submit your online payment information. Then wait patiently by the front porch.

Finding a 3D Printing Service

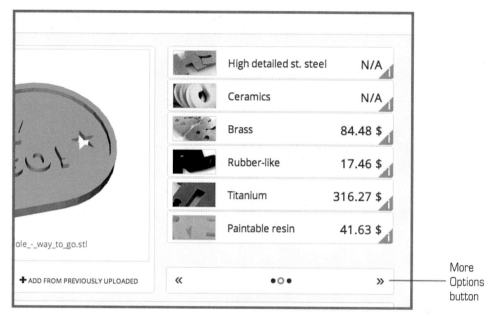

FIGURE 11.5 Even more material options are available from i.materialize.

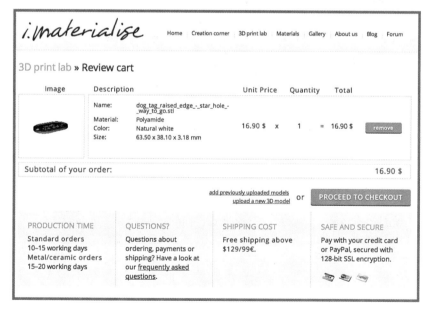

FIGURE 11.6 Check the details before placing your order.

CHAPTER 11: Expanding Tinkercad's Usefulness

Soon enough, your order will arrive, and you'll be able to hold your model in your own hands. Figure 11.7 shows a dog tag I ordered for a special occasion for my son.

FIGURE 11.7 Hold something you designed in Tinkercad in your hands.

> **NOTE**
>
> Remember, i.materialize isn't the only 3D printing service out there, but I do like its simple interface and the fact that I can immediately order my item without having to create any kind of user account. (You can create an account if you wish to track all your model orders, but it's not required.) The three other 3D printing services do require you to create an account before you begin ordering models from them.

As you can see, ordering a 3D model online is fairly simple. If you want to use a 3D printing service that's not a Tinkercad partner, you can simply download the STL file to your computer and be prepared at some point to upload that STL file to your chosen 3D printing service.

Finally, you may be lucky enough to live in a town or city where 3D printing services are starting to appear. If that's the case, you can save the cost of shipping if you're willing to pick up your model. With a local 3D printing service, you're likely to follow the same procedure as with an online-only one: Upload your file to the company's website and make your configuration selections. It's also possible that the company will ask you to just email your model file, along with specifications such as size, color, and material. Just do a Google search for "3D printing services" plus the name of your city to see if any 3D printing services exist locally.

Taking Your Object into *Minecraft*

You're probably familiar with *Minecraft* (https://minecraft.net), a video game from Mojang that's available on many platforms, including computers, tablets, and game consoles. Players create worlds of their own by "mining" blocks (cubes) of different materials to build structures, vehicles, weapons, and much more.

Figure 11.8 shows the first-person view from my son's custom world, which has trees and grass and even a lake. Notice how blocky the game is; part of the appeal of Minecraft is the simplicity of the graphics and the game play.

FIGURE 11.8 Welcome to a world in *Minecraft*.

CHAPTER 11: Expanding Tinkercad's Usefulness

One of the primary tasks in Minecraft is to mine...to dig. Another is to build. You discover different types of materials and make plans to use the materials to build items such as tools and weapons. In the scene shown in Figure 11.9, I've dug a hole and I've built a nice straight wall from the materials I discovered in the ground.

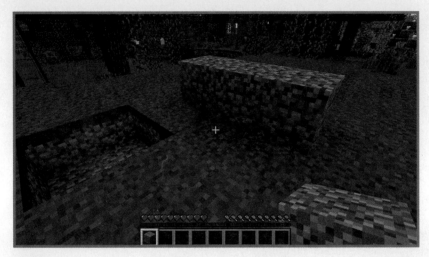

FIGURE 11.9 Digging and building are two constant tasks in *Minecraft*.

If you've never played *Minecraft* before, you can explore its rules and goals on your own. You may find that *Minecraft* can be quite addictive once you begin creating buildings and other objects in your world.

Why am I talking about *Minecraft* here, in a book on Tinkercad? Because Tinkercad offers you the ability to very quickly add items to your *Minecraft* world...no digging required.

Say that you want to add Abraham Lincoln wearing a Viking helmet to your *Minecraft* world. (Remember creating that in Chapter 10, "Where Can You Find Existing Models?") Open that file in Tinkercad and then click the Design menu, followed by the Download for Minecraft option, as shown in Figure 11.10.

Tinkercad asks you to size the object for use in a *Minecraft* world. Keep in mind that one block in *Minecraft* is 1mm. So your 42mm (4.2cm) object in Tinkercad will stand 42 blocks tall in the sky when you import it into Minecraft. If you set your design size to 2, the Lincoln statue will soar to 84 blocks high, basically heading into the clouds. Leave the default setting at 1mm, as shown in Figure 11.11.

Taking Your Object into *Minecraft*

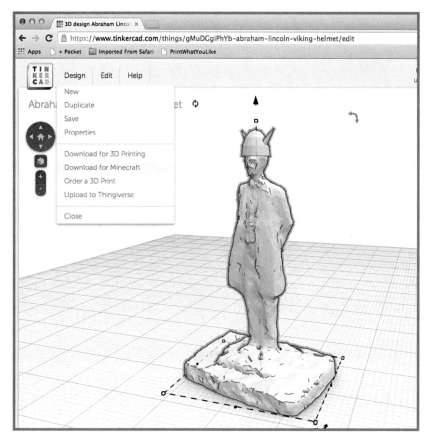

FIGURE 11.10 Converting a model into a *Minecraft*-friendly object.

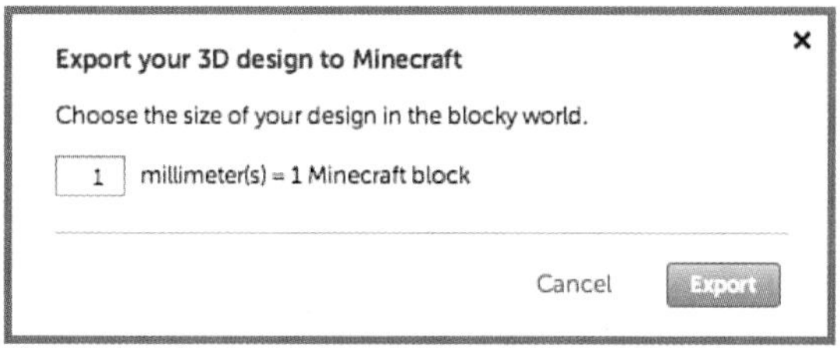

FIGURE 11.11 Select the size of your *Minecraft* world.

Click the Export button, and Tinkercad saves the file to your computer as a schematic file, as shown in Figure 11.12.

CHAPTER 11: Expanding Tinkercad's Usefulness

FIGURE 11.12 This file needs to be placed on a computer running *Minecraft*.

To import your object into *Minecraft*, you need a tool called MCEdit. It's a free download for Windows, Mac, and Linux computers. You can find it at www.mcedit.net.

You cannot have *Minecraft* open at the same time as MCEdit, or it could damage any work you've done on your world. However, be certain you have a world already created before you shut down *Minecraft*. If you don't already, go ahead and create one now and then shut down *Minecraft*.

Once *Minecraft* is closed, download the MCEdit tool for your operating system and run it. You should see a screen like the one in Figure 11.13. (My son has created a world called D's World. Your screen will have the name of your world in place of "D's World" on this screen.)

FIGURE 11.13 The MCEdit tool opens.

Click the world to which you want to import your Tinkercad model. The world opens up in MCEdit and looks a little like what you see in *Minecraft* but with option buttons running along the bottom of the screen. Use the W, A, S, and D keys and mouse to move up into the sky so you can look down on your island, as shown in Figure 11.14.

FIGURE 11.14 Your *Minecraft* world opens in MCEdit.

Click the Import button and browse to the folder or location where your Tinkercad object file (the schematic file) is located. Open it and use your mouse pointer to pick the location on the world where you want to place the object. Figure 11.15 shows that I'm going to place the Lincoln statue in a nice, visible spot.

When you're happy with the placement of your object, click the Import Lock button on the left side of the screen to finalize the location. Click the MCEdit menu at the top and choose Save and then click the menu again and choose Quit, as shown in Figure 11.16.

CHAPTER 11: Expanding Tinkercad's Usefulness

MCEdit menu

Import lock button

FIGURE 11.15 Place your object in the world.

FIGURE 11.16 Save and close MCEdit before you run *Minecraft* again.

After MCEdit closes, open *Minecraft* and the world you selected in MCEdit. Look around. You probably won't have any difficulty locating the model, as you can see in Figure 11.17.

FIGURE 11.17 Go on a hunt for your imported model.

NOTE

You might notice that the imported model is extremely blocky. The model of Lincoln wearing the Viking helmet wasn't very detailed to begin with, and you'll definitely have a less blocky appearance with more detailed objects. But *Minecraft* creates objects from cubes, so objects will always have that *Minecraft* look to them, no matter how detailed the original model.

If you have friends who play *Minecraft*, you can easily pull in some amazing objects such as castles, aircraft, and even a giant statue of yourself. Yes, yourself!

In Chapter 12, "Special App for Turning Real-World Objects into 3D Models," I'm going to show you a tool that you can use to create a 3D model of yourself, and you're going to have some fun with it as you finish one final project.

12

Special App for Turning Real-World Objects into 3D Models

In This Chapter
- Converting real objects to digital models
- Improving your 3D modeling skills

Throughout this book, you've learned how to use Tinkercad to create your own 3D models and how to turn those models into physical objects you can hold in your hand. Tinkercad provides a number of tools and features that you are now ready to use as you push your model-building skills to create even more advanced objects.

You've learned how to use the various shapes that Tinkercad provides; the Mirror, Align, and Group tools; the ability to import hand sketches; and many other Tinkercad features and abilities. You already have plenty to keep you busy creating, but before we wrap up the book, let's look at one more activity that I think you're going to enjoy: converting real objects into digital models, which is the opposite of what you've been doing so far.

Converting Real Objects to Digital Models

Figure 12.1 shows one of the latest releases from MakerBot, called the MakerBot Digitizer.

Basically, you place a small object on the revolving plate, and a laser scans the shape of the object to identify thousands of points on the object that it then uses to create a digital model in different file formats, including STL. This handy little device gives you the ability to scan an object and then make a duplicate of it with a 3D printer or printing service.

Other object scanners work in a similar manner, but right now the price of these types of digital scanners is around $600—too high for many hobbyist workshops and classrooms.

CHAPTER 12: Special App for Turning Real-World Objects into 3D Models

FIGURE 12.1 The MakerBot Digitizer connects to your computer.

Fortunately, there's another way to convert a toy or other object to a digital file that can be imported into Tinkercad, and it's free to use. All it requires is a camera capable of taking digital photos and some patience on your part.

123D Catch is a free app that's available for Windows PCs, iPhones, and iPads; there's also an online version that runs from a web browser. This app is another offering from Autodesk, the owners of Tinkercad. Here's how you use it:

1. Take a series of photos of an object from different angles. 123D Catch recommends taking 20 to 70 photos.
2. Upload the photos to Autodesk, using the 123D Catch app. Autodesk's special software examines the photos and attempts to "stitch" together a 3D representation of the object (and surrounding area) you photographed.
3. View the 3D file, rotating and zooming in and out as needed. Use the editing tools to crop/cut as well as fill in and repair holes in the 3D object.
4. After you have edited the object as desired, export it as an STL file that can be imported into Tinkercad and used however you wish.

The 123D Catch app is easy to use, but you will probably discover that using it well takes some practice. You need to learn through trial and error how many photos work best for creating a 3D object, and you have to experiment with editing to determine how best to create a detailed object model for use with Tinkercad. I've created a lot of 3D objects using 123D Catch, and I'm still learning new ways to use the tool and the editing features to get great 3D models.

Converting Real Objects to Digital Models

This chapter walks through the process of using 123D Catch so you can see what's involved. The screen shots in this chapter show the browser-based version running on my laptop. You can use the iPhone/iPad or Windows PC version instead if you like; the end result will be the same.

Taking the Photos

To get the 123D Catch app, point a web browser to www.123dapp.com. When you visit the website, you're greeted with a screen like the one in Figure 12.2.

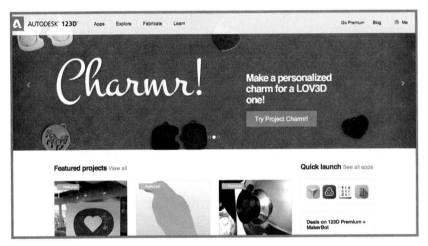

FIGURE 12.2 Start at the official 123D app website.

Click the Apps menu near the top of the screen and then click the 123D Catch option, as shown in Figure 12.3.

Scroll down the resulting page a bit, and you see three links for the Windows PC version, the version for iPhone and iPad, and the browser-based version of 123D Catch (see Figure 12.4).

Below the links for the three versions of the app are links to pages that introduce you to using 123D Catch. Watching the tutorial videos will greatly increase your skill with the app and help you obtain much better results with your scanned objects, so don't skip them! If you still have questions after watching the videos, you can click the Forum link to visit the 123D Apps Forum, where you can post your questions and get answers from other users and Autodesk staff members.

CHAPTER 12: Special App for Turning Real-World Objects into 3D Models

FIGURE 12.3 Select the 123D Catch app to view options.

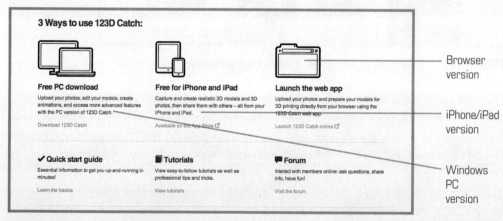

FIGURE 12.4 Choose the version of 123D Catch you want to use.

If you click the 123D Catch browser version link, you see a screen like the one in Figure 12.5. You can read tips, play with an example of a 3D object, and view other creations from other users. In this case, you're going to start by creating a new project by clicking the Start a New Project button.

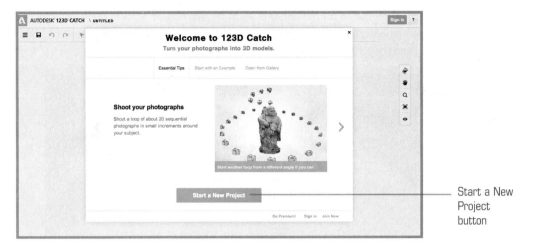

FIGURE 12.5 Open 123D Catch and start a new project.

A window like the one in Figure 12.6 appears, asking you to upload 6 to 70 photos. To grab these photos, you need to have them on your computer (your iPhone/iPad if you're using the mobile version) so you can easily browse to them.

FIGURE 12.6 Browse to the photos stored on your computer.

What if you don't have any photos of a 3D object? If that's the case, it's time to grab your mobile phone (if it has a camera) or a digital camera and take some pictures. I'm not going to go into a lot of detail here because the 123D Catch website's tutorials offer plenty of

CHAPTER 12: Special App for Turning Real-World Objects into 3D Models

advice on how to do this, but in a nutshell, you want to place your object where you can easily circle it as you shoot photos. Figure 12.7 shows a simple plastic robot I've placed on the table and am preparing to shoot.

FIGURE 12.7 Place your object on a table or other surface.

You may notice that I've placed some slips of paper with letters and little sketches on them. The software uses these as reference points to help with stitching the photos together. Again, the tutorials explain how these help and why you should use them.

> **TIP**
>
> When placing your object, consider that a textured background works better than a plain one, and a dull finish works better than a shiny one.

All in all, I took 63 photos, 10 of those were from slightly above looking down. Figure 12.8 shows a few of my photos from various angles.

Converting Real Objects to Digital Models

FIGURE 12.8 You need more photos to create a 3D model.

These are some of the photos I will use to try to create a 3D rendering of this figure. The next step is to click the Select Photos button on the screen shown in Figure 12.6. After 123D Catch imports the photos, it displays them as shown in Figure 12.9.

> **NOTE**
>
> 123D Catch may ask you to log in with your Autodesk user account. This is not the same as your Tinkercad account, so create one if you need to do so.

Next, click the Process Capture button and wait. You may have to wait a while as your photos are uploaded to Autodesk so that special software can examine them and attempt to re-create the photographed object as a 3D model onscreen. You should see a progress bar like the one in Figure 12.10 for each photo uploaded.

CHAPTER 12: Special App for Turning Real-World Objects into 3D Models

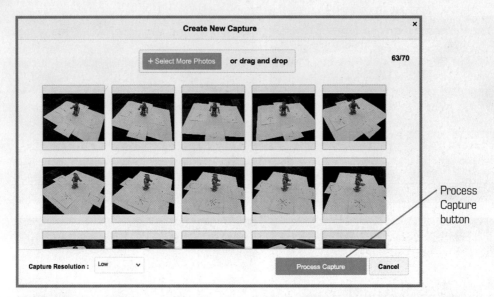

FIGURE 12.9 123D Catch imports the photos before beginning the "stitching" process.

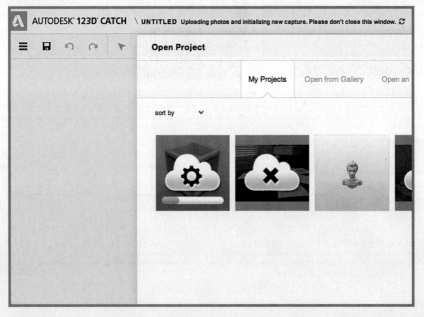

FIGURE 12.10 Files are uploaded, and Autodesk processes them.

NOTE

You should be aware that once your photos are uploaded to Autodesk, it can take between a few hours to a full day for your 3D model to be created. You'll receive an email (or a notification on your mobile device) when the 3D model is finished. Paid account users get priority, but users with free accounts will eventually have their models completed—patience is a virtue!

When 123D Catch finishes its process, you'll hopefully find a new 3D model listed in your Models list, as shown in Figure 12.11.

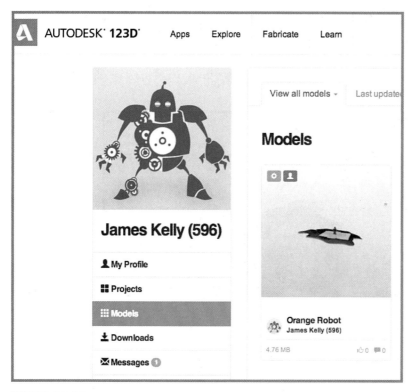

FIGURE 12.11 Your new object is listed in the Models list.

Click the new model to open its information screen (shown in Figure 12.12) and then click the Edit/Download button and select the 123D Catch option from the drop-down menu.

Your model begins loading. This can take some additional time, so be patient. When your model is ready, it opens on what is referred to as the Canvas, not the workspace (see Figure 12.13). You need to name the file, add some tags, and select a category for it.

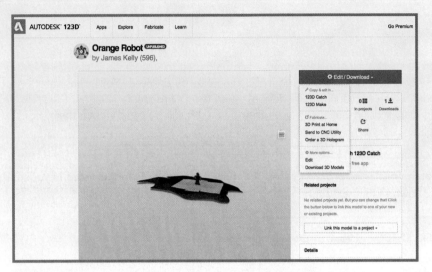

FIGURE 12.12 Your model's information screen.

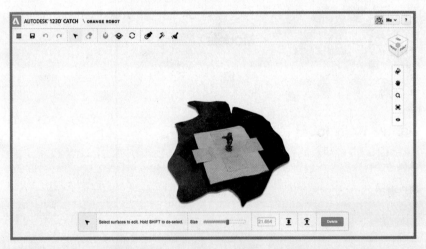

FIGURE 12.13 Your model is on the Canvas and ready for editing.

In Figure 12.13, notice that the 123D Catch app shows not just the robot but also the pieces of the paper surrounding it and some of the table as well. Don't worry; you can remove all this extra stuff and keep only what you want to save and import into Tinkercad. But you need to do a bit of editing of the 3D file in 123D Catch before you can use it in Tinkercad. Then, after you do some cleanup work, you need to export the object to an STL file. Finally, you can import it into Tinkercad.

Editing Your 3D Model in 123D Catch

There are a lot of things you can do to your 3D model within the 123D Catch app. This chapter focuses on converting a real object to an STL file that you can use in Tinkercad, but in this section I give you a little peek at the tools for cleaning up your model so that Tinkercad can easily and safely use it.

Before you begin editing your model, you need to get familiar with the Rotate and Drag tools. The Orbit tool allows you to move around, under, and above your model to view it from different angles. The Pan tool allows you to move the object (without rotating it) on the screen so you can better use the editing tools. Finally, you can always zoom in and out on your object. (If you're using a mouse with a scroll wheel on top, you can use that wheel to zoom in and out.) Figure 12.14 shows these tools.

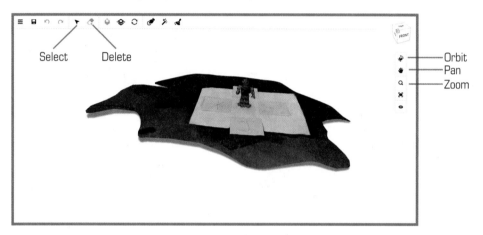

FIGURE 12.14 The basic model manipulation tools: Orbit, Pan, and Zoom.

The most important editing tasks you need to do with your model are remove unwanted material and repair/fill any holes. Unwanted material may include the surface on which your object is resting as well as any small slips of paper you may have placed around the object to assist with the stitching operation.

Fortunately, erasing unwanted sections of the 3D model is easy to do; you simply need to drag your mouse pointer around a section that you want removed. With your model open in 123D Catch, click the Select button. To get rid of as much of the surrounding material as possible without cutting away any of the actual surface area of the model, drag a line around the material you want deleted.

When you have selected an area, it turns color (a medium brown) so you can see the area you've selected, as shown in Figure 12.15.

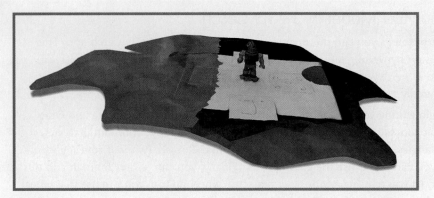

FIGURE 12.15 Check to make sure the selected area is correct for removal.

Click the Delete button, indicated in Figure 12.14, and the selected area is removed, as you can see in Figure 12.16.

FIGURE 12.16 Cleaning up the unwanted bits of the model is easy.

Continue to remove unwanted material and use the Zoom, Orbit, and Pan buttons to better see all the surrounding material that you want to delete. Zoom is very helpful for getting closer and closer to what will be the final model.

Converting Real Objects to Digital Models

> **NOTE**
>
> If there's a small amount of material that is just too difficult or tricky to remove, just leave it. You can use a hole object in Tinkercad to carefully remove the unwanted sections.

When you're done editing and deleting material, you should end up with a model that you can rotate around, just as you would in Tinkercad. Figure 12.17 shows that I've edited down the model so that only the figure remains, plus a little bit of the surface it was sitting on.

FIGURE 12.17 Zoom and Orbit (rotate) around the object for better editing.

Now you need to save your model as an STL file that can be used in Tinkercad.

This model is only 3MB—not even close to Tinkercad's 25MB import limit. In this case, I'm going to reduce the file size even further because I am only interested in the top half of the figure. I used one tool called Plane Cut to cut off the bottom half, as shown in Figure 12.18.

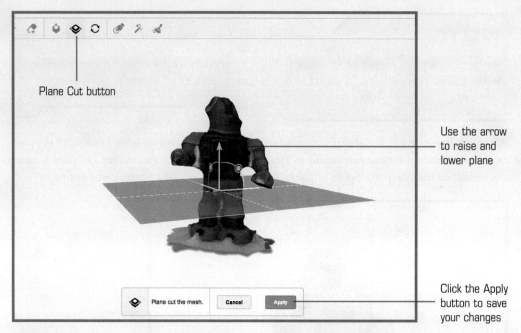

FIGURE 12.18 The final model is almost ready to open in Tinkercad.

Figure 12.19 shows the final half of the figure—the half that I plan to import into Tinkercad as an STL file.

FIGURE 12.19 You can use the Plane Cut tool to make fast and easy trims.

Saving Your STL File

Save your model in 123D Catch by clicking the Save button. Then click the Main menu (three horizontal bars) and click Export STL. The model is converted to an STL file and saved to your computer, as shown in Figure 12.20.

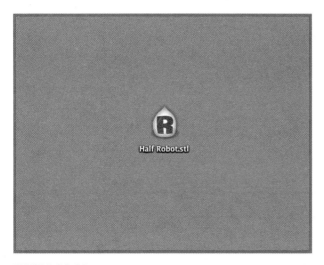

FIGURE 12.20 Save the STL file to your computer so you can import it into Tinkercad.

Import the file into Tinkercad (refer to Chapter 9, "More Useful Tricks with Tinkercad," for details) and wait for it to appear. You may have to increase the Scale to 300–500% to view smaller items you've photographed. One of the first things you'll notice about an STL file you import from 123D Catch is that it's not nearly as pretty as the original. As you can see in Figure 12.21, the model is now missing all the color and facial expressions, and the shape is a bit uneven.

Okay, let's be honest: The imported model isn't pretty at all. But I could have gotten a better result if I'd taken more photographs at a higher resolution. The 123D Catch application allows for up to 70 images, and the more you provide, the better the final stitched model will look.

Even though the model I have in Tinkercad now isn't pretty, I'm okay with it because I can use the built-in Tinkercad shapes (and hole objects created from those shapes) to smooth and cut away bad sections.

What can I do with this half of the figure? All sorts of things! For example, take a look at Figure 12.22, and you'll see that I'm converting this figure to sit on top of a set of strange robotic legs.

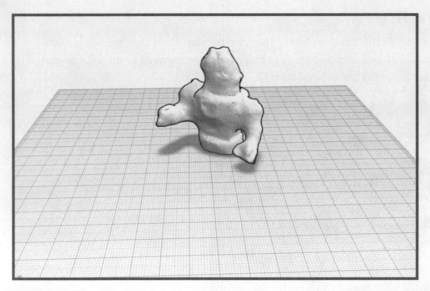

FIGURE 12.21 The imported object may not look exactly like the original.

FIGURE 12.22 Mix and match tops and bottoms to create custom toys.

Another way you could use this half of the figure would be to add it to a scan of a toy car or tank. You can easily take photos of your favorite toys and import them into Tinkercad to mix and match with designs you created in Tinkercad (such as the robot legs) or with other scans.

If you can sit still long enough to have 70 photos taken, you could even convert yourself to a model and import it into Tinkercad to create a statue of yourself. Or you could drop your head, arms, and chest onto another model to create some strange mixes.

Whatever you choose to photograph and create with the 123D Catch app, you can definitely have some fun by importing it into Tinkercad and modifying it however you like. When it's done, print it on a 3D printer or send it to a printing service, and you'll have a one-of-a-kind toy.

Improving Your 3D Modeling Skills

We've reached the end of the book—but not the end of your explorations. In addition to Tinkercad and 123D Catch, you'll also find additional free apps from Autodesk over at 123dapp.com. I highly encourage you to read the information on those free apps and see how you might integrate them into your Tinkercad designs.

If you're looking to move from Tinkercad to a CAD application that has more features and tools, you won't have to spend a dime. Autodesk makes 123D Design (you can find an overview of the app in Appendix C, "A Closer Look at 123D Design") available for both Mac and Windows PCs, and it's a perfect CAD upgrade that builds on what you know about Tinkercad with more tools. You'll find its interface very easy to use, and you'll quickly discover some of the features and tools that 123D Design and Tinkercad share.

If you enjoy creating 3D models, you're going to want to keep learning and experimenting with other CAD applications. You'll find a list of other CAD applications in Appendix A, "More Free CAD Applications to Explore." Explore that list if you're looking for new challenges.

I hope you've enjoyed learning how to use Tinkercad (and additional services that work well with it). Remember that getting hands-on is the best way to learn and retain what you've learned, so open up Tinkercad and go design something. Add parts. Group items. Mirror objects. Import some hand sketches. Come up with your own creations and share them with family and friends. Smile wide because you've mastered Tinkercad and are ready to go!

Have fun!

More Free CAD Applications to Explore

Tinkercad has plenty to offer beginners in terms of CAD features and tools. Its simple interface and limited tool set are unintimidating and easy to learn. And because Tinkercad is available via a web browser, it doesn't require any special configuration or have any unusual installation requirements.

While Tinkercad would be placed solidly on a list of beginning CAD applications, anyone experienced with CAD can easily and quickly begin to use it for 3D modeling. Because Tinkercad has the ability to import SVG and STL files, users can import models from other users or even from their own hand sketches.

Still, there are size limitations for file imports, and Tinkercad frequently requires placement of parts using drag-and-drop techniques that put a lot of the responsibility for the accuracy of the placement on the user. Tinkercad can only export to STL files or *Minecraft* objects (for now), so taking your model into other applications will work only if that software imports STL 3D model files.

So where do you go when you've pushed the limits of Tinkercad and are looking for more power, features, and tools? The answer is a more powerful CAD application. There is no shortage of CAD applications out there; although most are not free, some are. The most powerful CAD applications have hefty price tags, but those are the types of programs that professional 3D model designers use in industries such as television/movies, consumer products, and architecture.

Thanks to the various shapes that Tinkercad provides; the Mirror, Align, and Group tools; the ability to import hand sketches; and the many other features and abilities, you've got plenty to keep you busy creating. At some point, though, you might want to expand your skills and need more tools for more advanced 3D designs. This appendix describes three CAD applications that are currently free. This list isn't comprehensive but is merely a launchpad for you to begin your investigation of free CAD software.

123D Design

The 123D Design app shown in Figure A.1 is a free CAD application from Autodesk (which also owns Tinkercad). It should be very easy for Tinkercad users to make the jump to 123D Design because it is similar to Tinkercad in many ways. It offers more tools and features (especially the Windows and Mac versions) than Tinkercad. (As a matter of fact, Appendix C, "A Closer Look at 123D Design," provides a more in-depth look at 123D Design and how it compares to Tinkercad.)

APPENDIX A: More Free CAD Applications to Explore

123D Design is available for Windows and Mac computers as a downloadable installation file; you can also get an online web-based version and a mobile version for iPhone and iPad. For more details on 123D Design, visit www.123dapp.com/design.

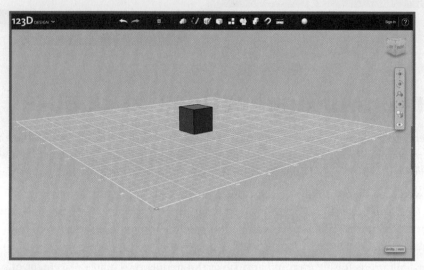

FIGURE A.1 The 123D Design application from Autodesk.

SketchUp

SketchUp could owe its popularity partially to being affiliated with Google. (Google purchased SketchUp and then sold it years later to Trimble.) But it's also partially due to just how simple SketchUp is to use. You install this application on your computer (Windows or Mac); you need a mouse or touchpad to add and manipulate objects. Users are fond of many of the shortcuts available, such as typing in a number while dragging a box and having the length of the object jump to that value immediately. An extension (that is, a downloadable add-on file) for SketchUp allows you to export objects as STL files.

There are two versions of SketchUp: a free version called SketchUp Make (shown in Figure A.2) and a paid version called SketchUp Pro. The price of SketchUp Pro depends on whether you're a student, a teacher or instructor, or a professional; for pricing information, go to www.sketchup.com and click Buy.

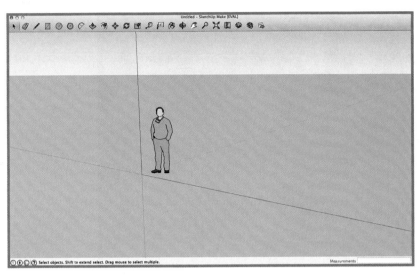

FIGURE A.2 The SketchUp application is available for Windows and Mac.

FreeCAD

FreeCAD is another popular (and free—hence the name) CAD application that offers users many more tools and features than Tinkercad. It is an open source application, so you might find modified versions of the original application, as well as a number of add-ons that have been created to perform specific tasks. It's available for Windows, Linux, and Mac computers, and it has an amazing amount of documentation and tutorials for beginners, as well as a forum with almost 3,000 users (see http://forum.freecadweb.org). FreeCAD (shown in Figure A.3) supports exporting to STL files, so users should have no trouble taking files from Tinkercad into FreeCAD or vice versa.

For more details on FreeCAD, visit www.freecadweb.org.

APPENDIX A: More Free CAD Applications to Explore

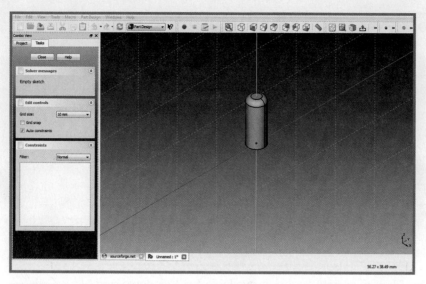

FIGURE A.3 The FreeCAD application can run on Windows, Mac, and Linux computers.

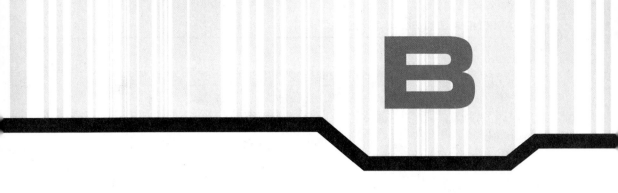

A Bonus Project

Wait, wasn't Chapter 12, "Special App for Turning Real-World Objects into 3D Models," supposed to be the final project? Yes, it was, but it's always fun to surprise people, and I want to pop a surprise project on you that I think you'll enjoy.

In this appendix, I'd like to show you how I've used Tinkercad on a more advanced project, along with a couple of the tools and techniques that you read about earlier in the book. With Tinkercad, sometimes you have all the tools you need right in the online application, and other times you have to do a little extra work before you open Tinkercad.

This bonus project starts with a story.

The Pinewood Derby

A few of my friends have young sons who are in Cub Scouts. Once a year, these dads and their young boys craft custom cars out of rectangular pieces of pinewood. Figure B.1 shows what that 1.75in. by 1.25in. by 7in. long block of wood looks like.

The event is called the Pinewood Derby, and it's been around a long time. How long? Figure B.2 shows my own Pinewood Derby racer from over 30 years ago.

During the event, a handful of racers place their handcrafted cars on a raised track. A switch is pulled, and the cars gain speed as they drop down a sharp hill, followed by a long straightaway to the finished line.

For most racers, the event isn't about winning. While attendees will see a wide range of cars that have been aerodynamically created to reduce drag and increase velocity on the hill, they'll also see a variety of cars that are made simply for the fun of creating something unique, funny, or eye-catching.

One of the main methods of creating a Pinewood Derby racer involves taping a set of templates over the wood block. Using a coping saw or a band saw, the father/son team carefully cuts along solid lines on the templates. Figure B.3 shows an example, with a Top template and a Side template taped to the sides of a block.

APPENDIX B: A Bonus Project

FIGURE B.1 This block of wood will be turned into a race car.

FIGURE B.2 My Pinewood Derby racer from the late 1970s.

The Pinewood Derby

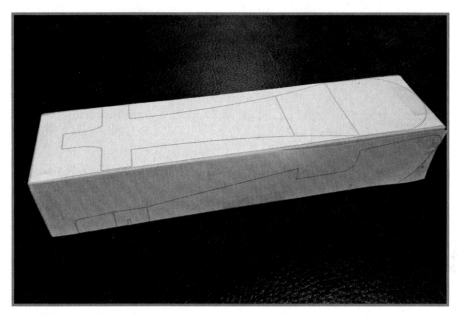

FIGURE B.3 Participants often use paper templates to guide their cuts on the block.

The team makes cuts on one side at a time. After all cuts are made, what should be left behind is a body shape. The car owners can then paint and decorate this shape any way they like.

Templates can be found all over the place online. Just search for "Pinewood Derby Templates," and you'll find hundreds of PDF files that you can download, print, and then tape to a block. The most common templates contain a Top pattern and a Side pattern that combine to create a themed car such as a NASCAR racer or an old-time hot rod. The Top and Side patterns must be used together to obtain the final desired shape.

The son of one of my friends emailed me recently because he knew that I use Tinkercad and a 3D printer to create physical objects from digital files. He asked me two questions:

- Can Tinkercad help me to see the final shape of a car before I make any cuts?
- Can Tinkercad help me create interesting gadgets to glue on to my car (like a fake engine or a pair of rocket engines)?

Fortunately, the answer to both questions is YES! In the following sections, I show how we used Tinkercad to test some shapes for a possible car before my young friend made any cuts to the wood. I'll also show you how we created a digital model of the car that we could use to test out custom gadgets to be printed by a 3D printer and attached to the final car.

Creating Digital Body Shapes

My friend's son had a great idea to mix and match various Top and Side templates. What would a surfboard-shaped Top template look like mixed with a Volkswagen Bug Side template? Or how about a Top template shaped like an hourglass mixed with a Side template shaped like a pickup truck? With each block of wood costing about $5, testing different mix-and-match patterns could get expensive. That's one of the benefits of building 3D models digitally: You don't have any materials costs, so you can happily make mistakes and try different things to your heart's content.

I asked this young designer if he could provide me with one Top template and one Side template that he'd like me to test. He sent me hand sketches of what he believed to be an eye-catching Top pattern for the racer and a Side pattern. He wasn't positive his combo would work because he couldn't visualize what the final shape would look like when the Top and Side patterns were applied to the block of wood.

I took his hand sketches (along with his measurements) and re-created them in a graphics program. I tackled the Top template first by drawing a rectangle to match the wood block's top dimensions of 1.75in. wide by 7in. long. A series of horizontal and vertical lines that you can see in Figure B.4 helped me connect the diagonal lines that represented certain cuts to be made.

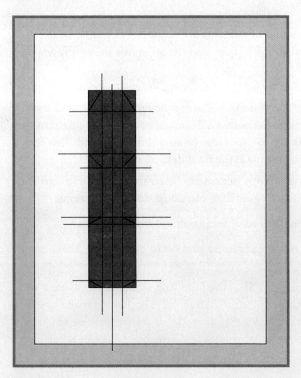

FIGURE B.4 I created the Top template in a graphics program.

Creating Digital Body Shapes

After developing the outline for the Top template, I filled it with color and deleted the extra lines to show the parts of the "block" that would not be cut away. What was left over is the solid shape shown in Figure B.5. I saved this as a JPEG image. (My graphics program doesn't allow me to save to the SVG file type required by Tinkercad.)

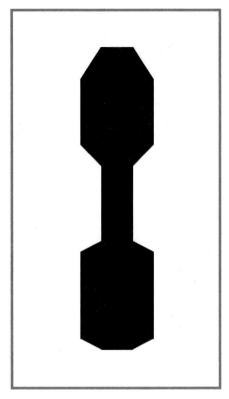

FIGURE B.5   The Top profile for the pinewood block, as a JPEG image.

Creating the Side template was just as easy. I created another rectangle with dimensions 1.25in. wide by 7in. long and then used another series of lines plus some circles to re-create the curves from the hand sketch. Figure B.6 shows this Side template, lined up above the Top template.

After I added the guidelines and circles, I used a Fill tool to fill in sections in black that would remain after the side cuts were made. Figure B.7 shows what the Side profile will look like in black; the gray-shaded areas and all the extra guidelines will need to be deleted.

APPENDIX B: A Bonus Project

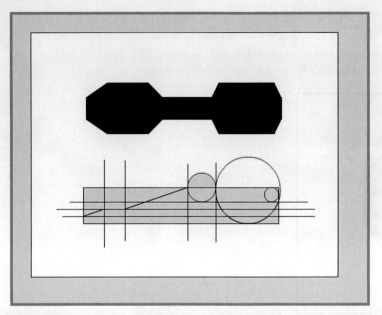

FIGURE B.6 The Side template begins to take shape with lines and circles.

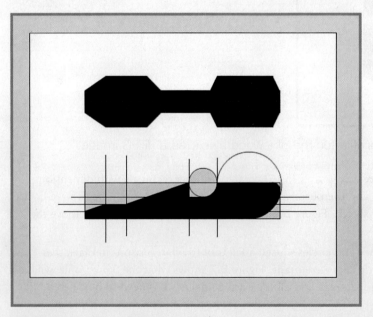

FIGURE B.7 The rough Side profile for the pinewood block, along with the completed Top template.

After I cleaned up the Side template, I was left with the shape shown in Figure B.8. I saved it as a JPEG file, too.

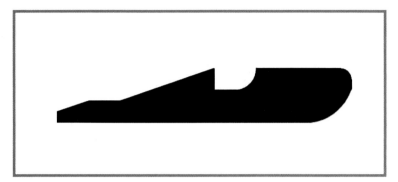

FIGURE B.8 The final Side template, saved as a JPEG image.

> **NOTE**
>
> I saved the two profiles as individual JPEG files. If I instead saved them together as one image and then converted that to an SVG file for importing into Tinkercad, the Tinkercad app would treat them as a permanently grouped object that cannot be ungrouped. That wouldn't be good.

Next, I converted each JPEG image to its own SVG file, using the www.online-convert.com website mentioned in Chapter 9, "More Useful Tricks with Tinkercad." I remembered to select the Monochrome option before clicking the Convert File button, as shown in Figure B.9.

The www.online-convert.com website created my two SVG files, as shown in Figure B.10. Because I created the two templates using the same scale in my graphics program (length of 7in., for example), I can be confident that they will import at the same scale.

> **NOTE**
>
> I'll skip the steps involved in importing these two images. If you need help with these steps, refer to Chapter 9.

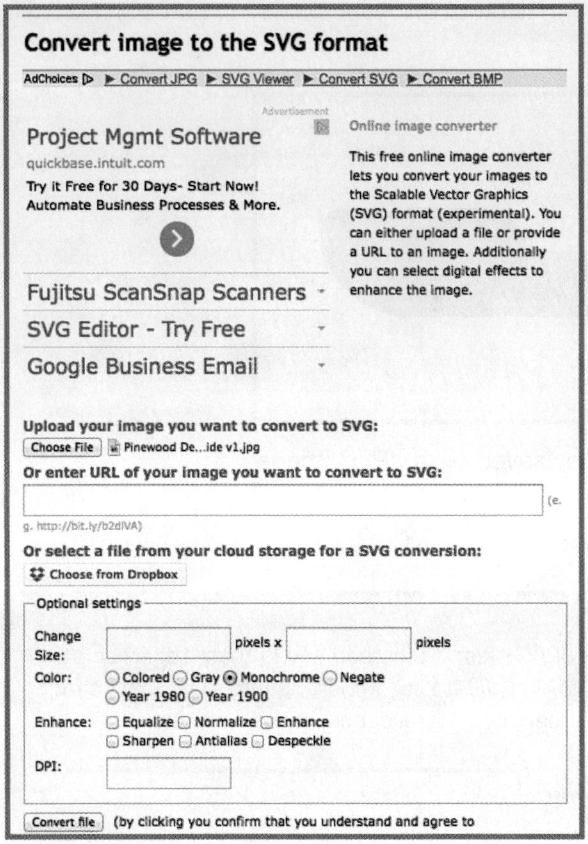

APPENDIX B: A Bonus Project

FIGURE B.9 Use www.online-convert.com to create the two SVG files.

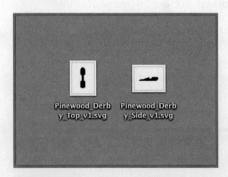

FIGURE B.10 Two SVG files, ready to be imported into Tinkercad.

When importing these SVG files, the objects are a bit larger than the workspace, as you can see in Figure B.11.

Creating Digital Body Shapes

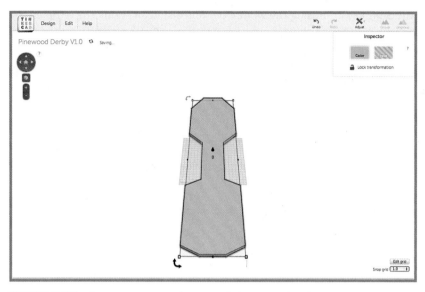

FIGURE B.11 Import a template and then resize it.

The problem is that Tinkercad has the current workspace set to use millimeters, not inches. To deal with this, I just click the Edit menu and then select Grid Properties; then I change the dimensions of the workspace to 10in. each, as shown in Figure B.12.

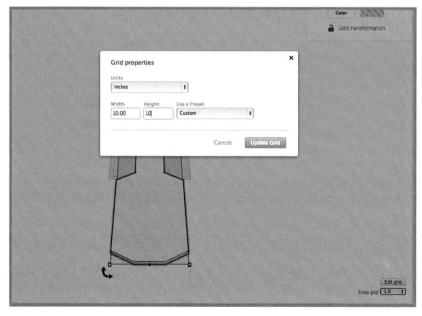

FIGURE B.12 Adjust the workspace dimensions to use inches.

By holding down the Shift key and using the white dot controls, I can shrink the template shown in Figure B.11 so that its length is as close to 7in. as possible. I also want to rotate the Top template 90 degrees. Figure B.13 shows the results of these changes.

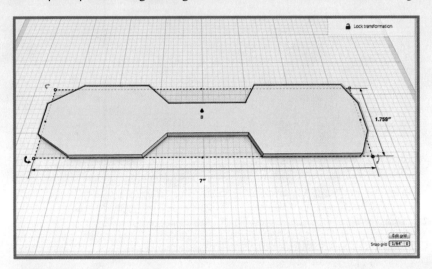

FIGURE B.13   Use the Shift key to shrink all dimensions with the same ratio.

There are two things I'd like you to notice in Figure B.13. First, the length is an exact 7in., but the width is 1.759in., not 1.75in. I could try to fix this by adjusting only the width of the template (dragging only one of the white dots to adjust only the width), but 0.009in. is close enough to the actual measurement for my purposes. Also, note that I changed the Snap Grid setting to 1/64in. so that I can adjust the dimensions in very tiny increments. This helped me get the length to an exact 7in. I could probably try to get the width to exactly 1.75in., but then the length would probably be slightly off.

Figure B.14 shows that I've imported the Side template, and it also needs to be reduced in size to have a length of 7in.

Once again, the Side template is forced to a length of exactly 7in., but its length is 1.263in., not the desired 1.25in. Again, this is close enough for me to work with, but if you like, you can always adjust a single dimension with a single white dot control while not holding down the Shift key. Just be certain if you do this that you don't adjust the length dimension.

Figure B.15 shows the resized Top and Side templates, imported and ready for creating a 3D model of this mix-and-match Pinewood Derby racer. What happens next?

Creating Digital Body Shapes

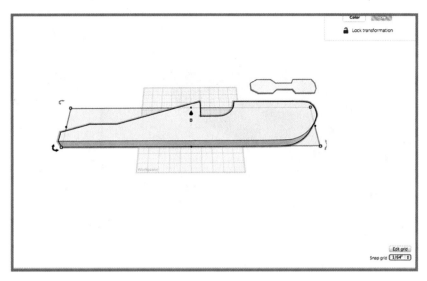

FIGURE B.14 Import the Side template and adjust its size.

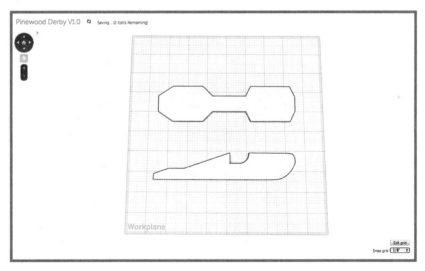

FIGURE B.15 The Top and Side templates are ready to be used to create a racer.

I knew I needed to start with some exact measurements, so I bought a pinewood block from a craft store and measured it. These dimensions were 1.75in. × 1.25in. × 7in. (width × length × height). Figure B.16 shows my virtual pinewood block.

APPENDIX B: A Bonus Project

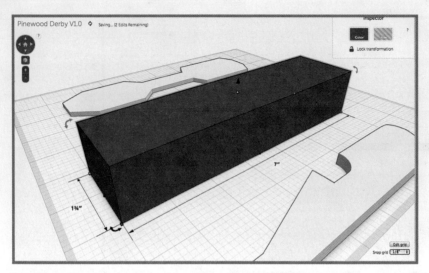

FIGURE B.16 Create a solid rectangle with the dimensions of the physical wood block.

I need to line up the patterns before I can place them over the block. If I change the Snap Grid setting to 1/8in. (by clicking on the Snap Grid button in the lower-right corner of the workplane), dragged objects will "snap" to the nearest 1/8in. grid square. I line up the back of the Top template, Side template, and block on the same grid line, as shown in Figure B.17.

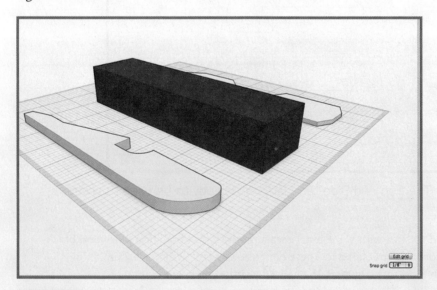

FIGURE B.17 Line up the templates and the block along a shared grid line.

"Carving" the Block

Pause here and look what I have. I have a solid block that represents the pinewood block that I want to "carve." I have a Top template that shows what solid materials need to be removed from the block when looking straight down (from the top) on the block. I also have a Side template that shows what solid material needs to be removed from the block when looking from the side. Now the fun begins.

"Carving" the Block

Based on your experience in this book, you may have guessed that I need to create a hole object of some sort to remove the unneeded material from the block. The way to do that is to create two copies of the block that will make up the top hole object and the side hole object. I'll start with the Top template.

Figure B.18 shows that I've spread things out a little bit so I have more working room. I've placed two copies of the block on the workspace as well. One block will remain as the "solid" material and the other two blocks will be turned into hole objects that will be merged and grouped with the "solid" block, leaving behind only the material that will make up the final car.

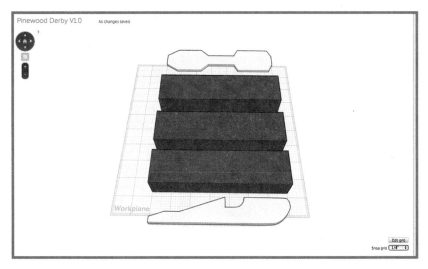

FIGURE B.18 Create two additional blocks to serve as hole objects.

I know the block is 1.25in. in height, so I increase the height of the Top template to 1.5in., as shown in Figure B.19. I made it slightly taller so when it is merged with the "solid" block, I can still see the outline provided by the templates as well as ensure that a full and complete hole is made through the solid block.

Next, I need to drag the Top template over one of the blocks so that it intersects the block and shares the same grid line (at the back). Figure B.20 shows that the Top template object and one of the blocks are now merged.

APPENDIX B: A Bonus Project

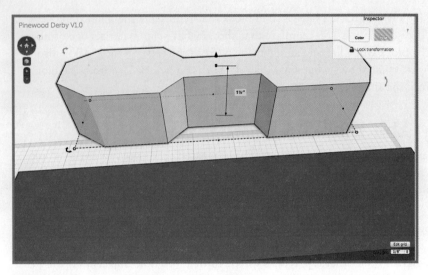

FIGURE B.19 Increase the height of the Top template.

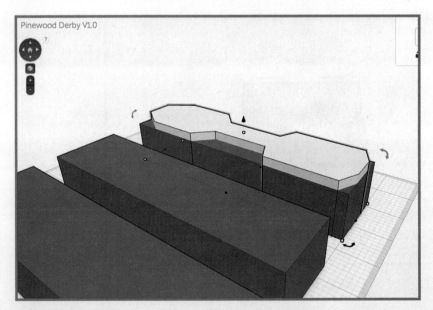

FIGURE B.20 Merge the Top template with a block.

After merging the Top template object and a block, I click just the Top template object, making certain only the Top template is selected and not the block with which it is merged. Then I click the Hole button, and parts of the block disappear, as shown in Figure B.21.

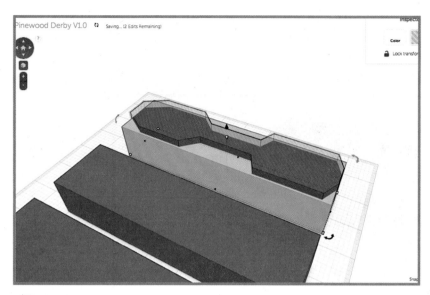

FIGURE B.21 Turn the Top template object into a hole object.

Finally, select both the Top template and the block and click the Group button. Wait a few seconds, and a hole that matches the shape of the Top template appears, as shown in Figure B.22.

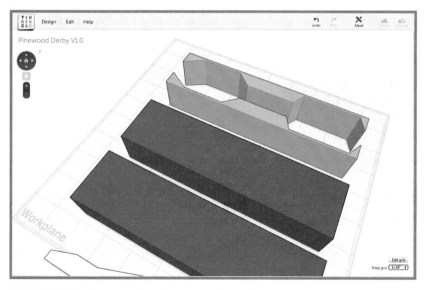

FIGURE B.22 Group the Top template object and block to create the final hole object.

I now have a Hole object that I'll use later to carve the digital block I designated as the "solid" block. But first I need to create a Side hole object. I do it in much the same way I made the Top hole object, but first I need to rotate the Side template so it's standing up, as shown in Figure B.23.

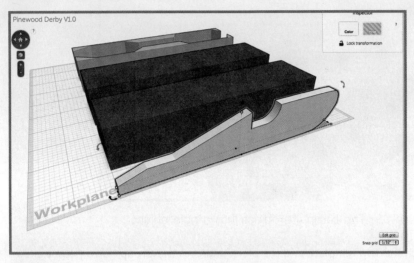

FIGURE B.23 Rotate the Side template up by 90 degrees so it's standing upright.

Next, I increased the width of the Side template so it was as close as possible to the 1.75in. width of the wood block. (The final measurement was 1.756 in. and was close enough for my purposes.) Figure B.24 shows that it's very close to the desired measurements.

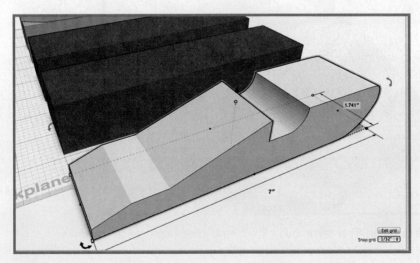

FIGURE B.24 Increase the Side template to match the block's width.

"Carving" the Block

Now I perform the same steps as with the Top template: Merge the Side template with the block, select the Side template and turn it into a hole object, and then group the Side template with the block. Figure B.25 shows what's left behind.

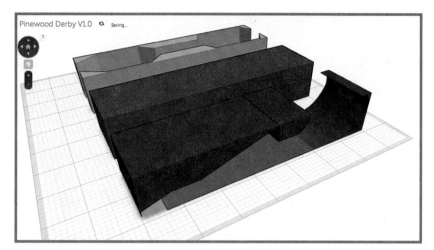

FIGURE B.25 Create the Side hole object.

NOTE

If you're wondering why I didn't just create one block and apply the Top and Side templates to it to create a single hole object, it's mainly due to experience working with Tinkercad. The more objects you group together and the more you use the Hole button to create holes, the more "sluggish" that object becomes to work with. And undoing any mistakes often requires using the Undo button frequently, forcing you to undo work that might have been correct in the first place. When in doubt, try to break up any hole objects you make into only two or three objects to be merged and grouped to avoid frustration down the line.

Can you guess the next step? That's right: I need to now turn both of the blocks with holes in them into hole objects and merge those with the final solid block. I'll start with the Side hole object.

I select the Side hole object and then click the Hole button to turn it into a new hole object. Figure B.26 shows the result.

APPENDIX B: A Bonus Project

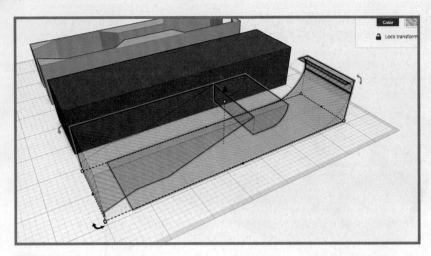

FIGURE B.26 Create a new Hole object to "carve" out the side of the block.

Drag this new Side hole object over the solid block to merge them. Select both and then click the Group button. Figure B.27 shows what's left over.

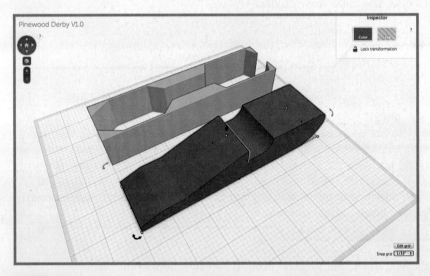

FIGURE B.27 The block with the Side hole object applied.

I'm almost there. Next, I select the Top hole object and turn it into a new hole object. Then I merge that with the remaining block, select both objects, and group them. I end up with the final shape of the test racer, shown in Figure B.28. It's definitely unique!

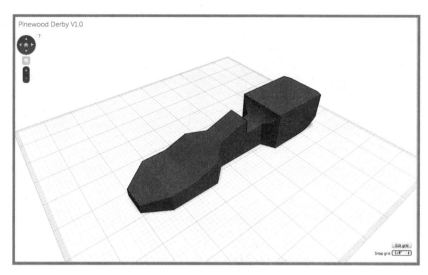

FIGURE B.28 The final shape of the Pinewood Derby digital racer.

All that's left is to add the wheels so we can see what the final racer will look like. The wheels are 1.25in. in diameter and 1/4in. thick, and although there are suggested places for them to be "nailed" into the original wooden block, race car designers are allowed to change the location.

Figure B.29 shows that I've created four wheels and placed them in their test locations.

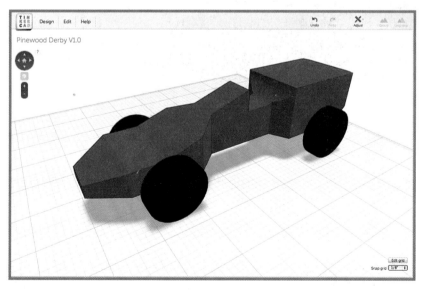

FIGURE B.29 The final shape of the Pinewood Derby digital racer.

APPENDIX B: A Bonus Project

I took some screenshots of the car from different angles and provided them to the young designer. He wants to continue to work on it, so I ended up exporting the design as an STL file that my friend can import and continue to work on.

By itself, the car looks very interesting. It's got a completely unique shape probably not seen in any race to date. Combine the shape with a custom paint job and some decals and a racing number, and my friend's son should have a car he can be proud of. But I wanted to help my young friend even more.

One of the things we had discussed is creating some small gadgets in Tinkercad and printing them out on a 3D printer. These would be glued to the car's body to give it character. Figure B.30 shows three gadgets (engine, left side tube, and a tube on the right that is hidden from view) I created in Tinkercad and placed on the digital car. If he likes these modifications, I'll send him the STL file, and he can print it on a 3D printer and add it to his racer.

FIGURE B.30 Every car needs an engine and side mufflers.

In this appendix, you've seen how I used real-world measurements (from the block and the young designer's hand sketches) to create a digital model. You could also use the reverse process: First design a car in Tinkercad and then take measurements from that model to create a set of templates from which to cut the car.

With many other projects as well, using Tinkercad to design as much as you can in the virtual space will often save you time and money.

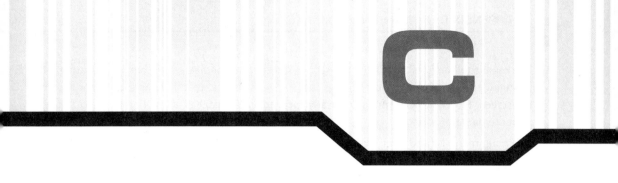

A Closer Look at 123D Design

Autodesk has an entire collection of CAD-related free applications for Mac and Windows, as well as a number of similar free apps for iPhone and iPad. You can find them at 123dapp.com. One of Autodesk's offerings is a 3D modeling application called 123D Design that is similar to Tinkercad but offers some additional tools and features.

This appendix introduces you to the 123D Design application. You'll learn about some of the features it shares with Tinkercad as well as some that are unique. The screenshots in this appendix were taken using the Mac version of the 123D Design application, but the Windows and iOS versions work the same way.

123D Design Interface

Figure C.1 shows the main 123D Design screen.

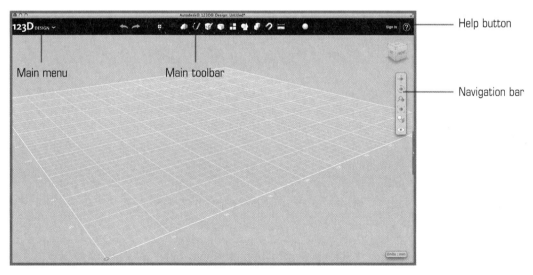

FIGURE C.1 The 123D Design user interface is simple and uncluttered.

APPENDIX C: A Closer Look at 123D Design

The main toolbar runs across the top of the screen, and the navigation bar (for rotating and changing the view, among other things) appears in a vertical rectangle near the right edge of the screen. Hovering your mouse over any toolbar button causes a small hint box to appear with the name of the tool.

The main menu is hidden at first, but a click on the 123D Design text in the upper-left corner of the screen causes the menu to appear, as shown in Figure C.2.

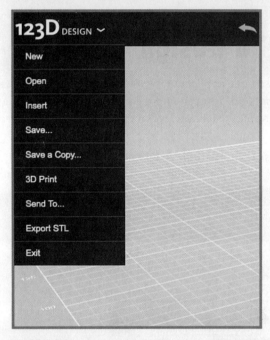

FIGURE C.2 The main menu for 123D Design offers many familiar options.

A quick look at the main menu shows many options that you should already understand. New, Open, Save, Save a Copy, Export STL, and Exit are all self-explanatory or make sense to you, based on your work with Tinkercad.

The Insert menu item functions as an import feature, allowing you to pull in other 123D Design objects from your library or from other users who have shared their designs to the 123D library. Unfortunately, the Insert menu item does not import STL files (but Autodesk will hopefully fix this in the future). 123D Design saves its files in its own format, .123dx, as shown in Figure C.3.

The 3D Print option in the main menu allows you to print directly to any 3D printer that is supported by the Autodesk 3D Print application—currently only MakerBot Replicator printers, Objet Connex 500, and Alaris 30 are supported, but this might change as Autodesk adds new 3D printers.

FIGURE C.3 123D Design saves files with the .123dx file type extension.

> **NOTE**
>
> You must install the Autodesk 3D Print application to print to supported 3D printers. You can also use this app to clean and repair a model before exporting it as an STL file.

The Send To option in the main menu allows you to send your model to other apps in the Autodesk 123D family, as well as choose 3D printing services to print your model.

You'll learn more about a few other items on the screen later in this appendix.

The Main Toolbar

You'll want to spend some time experimenting with all the various tools that appear on the main toolbar. This section gives you a quick overview of what most of them do.

> **TIP**
>
> Click the Help button for more details on how to use all the tools and features in 123D Design.

APPENDIX C: A Closer Look at 123D Design

Figure C.4 shows the complete main toolbar.

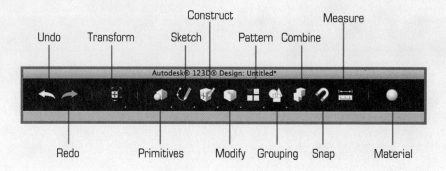

FIGURE C.4 The main toolbar in 123D Design.

In some instances, moving your mouse pointer over a toolbar button opens a mini-toolbar. For example, Figure C.5 shows the mini-toolbar that appears below the Primitives button.

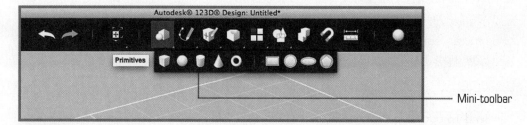

FIGURE C.5 Mini-toolbars appear when you move the mouse pointer over some buttons.

One of the simplest operations you learned to do in Tinkercad is to place a primitive on the workspace. You can place the primitive shapes in 123D Design's Primitives mini-toolbar on the workspace by simply clicking a shape (such as a cube) and moving your mouse pointer to the location where you want to drop it. A small dot appears inside the shape of the primitive, as shown in Figure C.6. Clicking where you want to place the primitive causes the dot to disappear and the solid object to be placed.

In addition to giving you access to primitive shapes, the Primitives mini-toolbar also allows you to draw outlines of shapes such as rectangles and circles. In addition to placing 3D objects on the workspace, you can place 2D shapes on the workspace, as shown in Figure C.7.

Don't worry if you're not happy with the placement or size of a 2D shape or 3D object on the workspace. Once you've placed 2D shapes and 3D objects on the workspace, you can easily modify their dimensions and locations by using what the 123D Design app refers to as selection-based options.

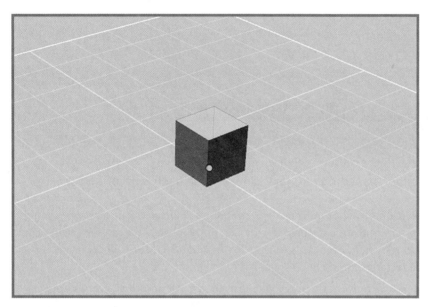

FIGURE C.6 Place a primitive on the workspace.

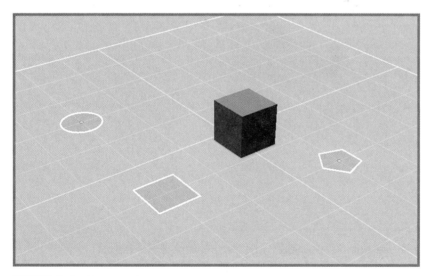

FIGURE C.7 Place 2D shapes on the workspace.

If you want to change the width and length (not height) of a rectangle 2D shape on the workspace, for example, click the rectangle, and a small gear appears nearby. Move your mouse pointer over the gear, and a small blue toolbar appears, like the one in Figure C.8.

APPENDIX C: A Closer Look at 123D Design

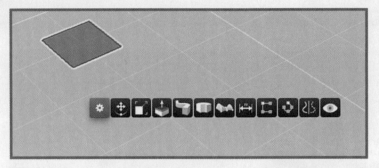

FIGURE C.8 Selection-based options are different for each type of object.

The rectangle's selection-based options toolbar contains 11 different buttons. Move your mouse pointer over any button, and the name of the tool appears. You can investigate each of these tools on your own. For fun, let's look at the first button, Move, which lets you shift the location of the rectangle on the workspace.

When you click the Move button, the rectangle turns light blue, and three arrows appear near the object, in addition to three white circles, as shown in Figure C.9. The arrows let you move the rectangle left/right, forward/backward, and up/down with respect to the flat workspace. Each of the white dots allows you to rotate the object on one of the three axes.

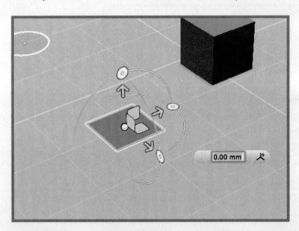

FIGURE C.9 The Move button lets you move and rotate a 2D object.

Figure C.10 shows that I've moved the rectangle closer to the "front" of the workspace (away from the 3D cube), and I've rotated it −56.9 degrees on the Z axis.

Moving and rotating a 3D object is slightly different from moving and rotating a 2D object because a 3D object has different faces. The cube, for example, has six faces (sides). If you click one of the sides, you get the familiar gear and selection-based options toolbar shown in Figure C.11.

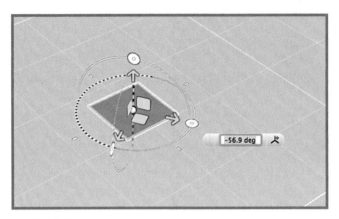

FIGURE C.10 The rectangle has been moved and rotated.

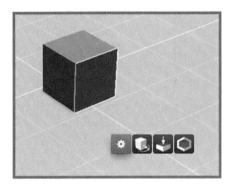

FIGURE C.11 A selected face and its selection-based options toolbar.

The problem here is that the selection-based options toolbar applies only to the selected face, not to the entire cube. Any of the tools you select here apply only to the face you clicked. If you want to manipulate the entire cube (for example, by moving or rotating it), you need to select the entire cube by drawing a selection window around it. After you do that, the 3D object's selection-based options toolbar appears near the bottom, as shown in Figure C.12.

To shrink or enlarge an object while keeping all the dimensions at the same ratio, you select the object and then click the Transform button on the main toolbar. Select the Scale option, and a small arrow appears inside the object, as shown in Figure C.13.

APPENDIX C: A Closer Look at 123D Design

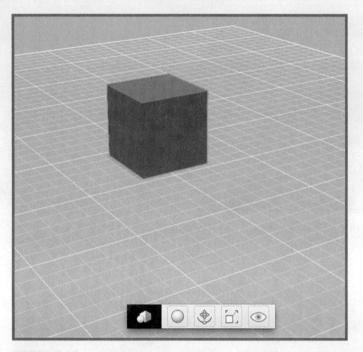

FIGURE C.12 Select an entire object to manipulate all sides/faces.

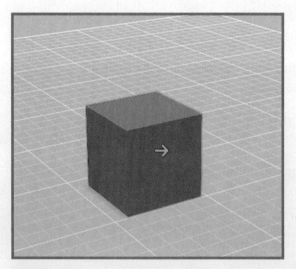

FIGURE C.13 Use the Scale button to shrink or enlarge an object.

The object shrinks with all dimensions (height, length, width) remaining to scale. This might cause the item to float above the work surface as its height decreases, as shown in Figure C.14.

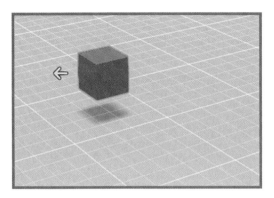

FIGURE C.14 The 3D cube shrunk with the ratio of the dimensions held constant.

Want to give a 2D rectangle shape some height? No problem. Click the rectangle and then from its selection-based options toolbar, choose the Extrude button (third from the left). A small up-pointing arrow appears, as shown in Figure C.15.

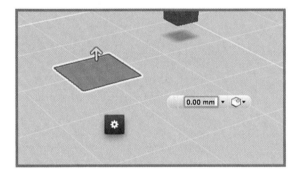

FIGURE C.15 Use the Extrude button to turn a 2D object into a 3D object.

Drag the arrow up and watch as the 2D rectangle turns into a 3D object. You can control the thickness/height of the rectangle with the small dimension text box shown in Figure C.16.

The Primitives button places preconfigured objects with fixed dimensions onscreen. You can select an object, click the Transform button, and then select Scale to resize an object. What if you want to add a 2D or 3D object and control all of its dimensions (length and width plus height for 3D) from the beginning? Click the Sketch button on the main toolbar and then select something like the Rectangle option.

As you can see in Figure C.17, the Sketch tool allows you to click (and hold) and drag your mouse pointer to define the length and width of the rectangle.

APPENDIX C: A Closer Look at 123D Design

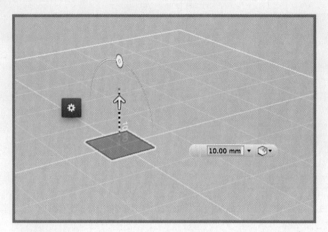

FIGURE C.16 Use the dimension box to control the 3D object's height.

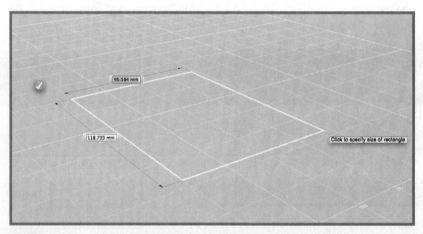

FIGURE C.17 Create a custom-sized object with the Sketch tool.

Once you have the width and length set, you can use the Extrude feature to give it height: Simply select the Rectangle button and then choose Extrude from the selection-based options toolbar.

There's more to see in the main toolbar, but first let's examine the navigation bar for a moment.

The Navigation Bar

Once you start placing objects on the workspace, you'll probably need to change views occasionally to work on objects from different angles. For the example, Figure C.18 shows that two objects—a cube and a sphere—have been dropped on the workspace a slight distance apart.

FIGURE C.18 Two objects are placed slightly apart on the workspace.

Figure C.19 shows the navigation bar.

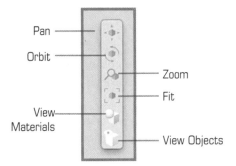

FIGURE C.19 The navigation bar and its various tools.

To get a bit closer to your models, click the Fit button. Clicking the Fit button forces the view to zoom in a bit while keeping all objects visible.

APPENDIX C: A Closer Look at 123D Design

Next, click the Pan button. The mouse pointer turns into a plus-sign icon. Now click and hold as you move the pointer onscreen. The Pan tool lets you shift the view left and right and up and down, without any rotation. Using it is an easy way to shift interest to an object while keeping one side (face) displayed. Figure C.20 shows that the view shifted so that the sphere is completely off the screen and the cube is the central focus. (You can click the Fit button again to have all objects once again displayed onscreen.)

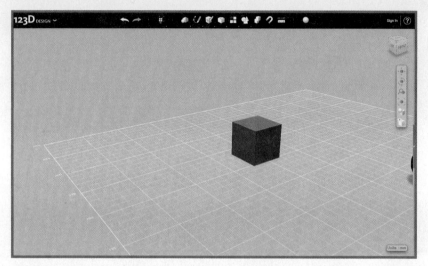

FIGURE C.20 Use Pan to shift the view without rotating any objects.

Zooming in and out is another simple task. Click the Zoom button, and your mouse pointer icon changes to a small vertical line with an arrow pointing up and another arrow pointing down. Click and hold to drag. Drag the mouse pointer up, and the screen zooms out. Drag the mouse pointer down, and the screen zooms in.

As you can see in Figure C.21, you can use the Pan button to center an object on the screen and then use the Zoom button to zoom in closer to do more detailed work on the object. A single click on the Fit button returns the screen to displaying all objects.

Finally, you can use the Orbit button to rotate around one or more objects on the screen. Click the Orbit button, and your circle surrounds the object onscreen, as shown in Figure C.22.

FIGURE C.21 Zoom in (or out) on an object with the Zoom tool.

FIGURE C.22 The Orbit button lets you rotate around the same face view.

If you leave the mouse pointer outside the circle, the mouse pointer icon looks like a small dot with a circular arrow going around it. (Unfortunately, screen shots do not pick up on this icon for some reason, so you'll need to try it to see the icon for yourself.)

If you click and hold while the mouse pointer is outside the circle surrounding the object, moving the pointer causes the view to change without changing the face you are observing. It's like rotating a virtual camera that's pointed at the object instead of rotating the object itself. The camera moves in a circle around the object while still pointing at the same face. Try it, and you'll understand the concept.

Move the mouse pointer inside the black circle, and you can rotate freely around the object by clicking and holding while moving the pointer around the screen. Again, this will make much more sense if you try it yourself. Figure C.23 shows that I've rotated the view so I'm looking up from underneath the cube.

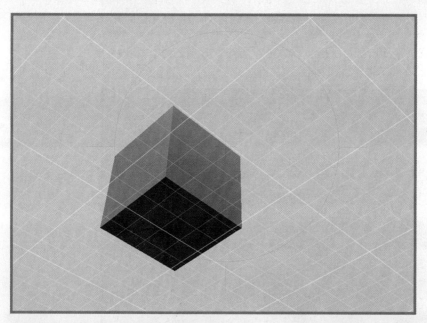

FIGURE C.23 The Orbit button also allows you to rotate freely.

There are two other buttons on the navigation bar that let you select materials for your objects and hide those materials (such as the color or texture) and flip back and forth between viewing dropped primitive objects (solids) and sketches (that you've created). Figure C.24 shows that the outlines of the two objects are the only things visible; this is helpful when you're trying to match up surfaces of two objects or view all the primitive and custom shapes that make up a larger, more complex object. Notice in Figure C.24 that there is now a custom circle sketch on the workspace that is also visible.

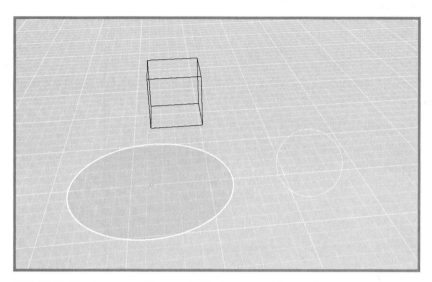

FIGURE C.24 View outlines of all objects without the solid colors and textures.

Figure C.25 shows what happens to this workspace when you choose to hide solids but leave sketches visible.

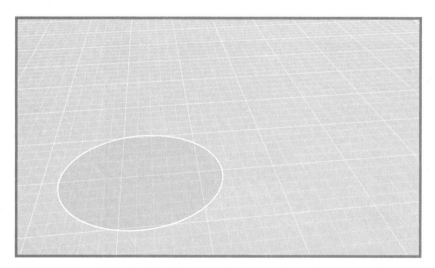

FIGURE C.25 Hide solid objects and view only custom sketches.

One final navigation tool that's helpful is View Cube, in the upper-right corner of the screen. Clicking the various faces of the cube lets you quickly change the viewing position of the selected object. If you click the top part of the View Cube tool, the view changes so that you're looking down on the three objects, as shown in Figure C.26.

APPENDIX C: A Closer Look at 123D Design

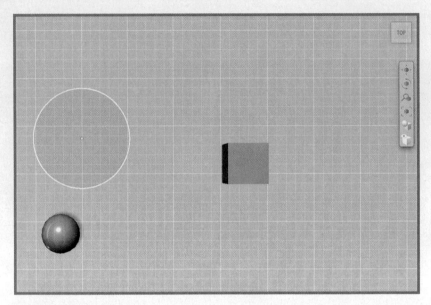

FIGURE C.26 Use the View Cube tool to quickly change viewing positions.

You can click the little icon of a house to return to a traditional angled view of all objects onscreen (called an *orthogonal view*), as shown in Figure C.27.

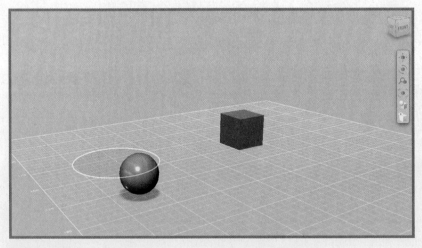

FIGURE C.27 Change back to an orthogonal view.

What's Left?

As with Tinkercad, with 123D Design you can group objects easily. You access the Group tool on the main toolbar. In addition, you can use the Ungroup button when selecting one or more objects, and you can use the Ungroup All button to ungroup everything on the workspace in one shot.

You can merge shapes together and use the Combine button to create a single solid object, as shown in Figure C.28.

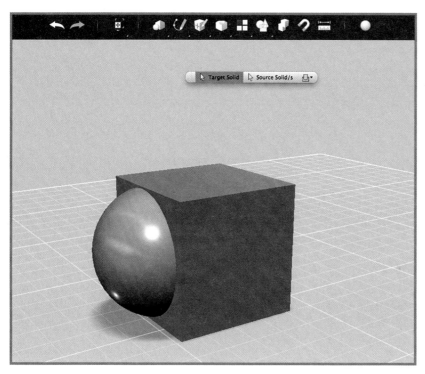

FIGURE C.28 Merge shapes together with the Combine button.

The basics of using 123D Design could fill an entire book. Even so, 123D Design is a basic 3D modeling app that shouldn't overwhelm you with too many tools and features. Consult the Help documentation, watch the tutorial videos, post questions on the forum, and experiment with each and every button you find until you better understand how the application can help you create even more advanced 3D models. A helpful book on 123D is *3D Printing with Autodesk: Create and Print 3D Objects with 123D, AutoCAD and Inventor* (by John Biehler and Bill Fane).

Index

Numbers
1D (one dimensional) points, 8
2D (two dimensional) objects, 8
3dhacker.com, 209
  3D modeling, definition of, 5-10
3D models. *See* **models**
3D printers
  controlling with printing software, 171-175
  costs, 156
  explained, 156-160
  fine-tuning, 174
  MakerBot Replicator 2, 159-160
  motors, 167-171
  nozzle movement, 167-171
  plastic filament, 165-167
  Printrbot Simple, 158
  RepRap Darwin, 158-159
  summary, 175-176
3D Printing: Build Your Own 3D Printer and Print Your Own 3D Objects **(Kelly), 156**
3D printing services, 212-217
3D Printing with Autodesk: Create and Print 3D Objects with 123D, AutoCAD and Inventor **(Bieher and Fane), 283**
123dapp.com/design, 244
123D Catch, 226-227
  editing 3D models, 235-238
  saving STL file, 239-241
  taking photos, 227-234
123D Design, 243-244, 267
  main toolbar, 269-270
    Combine button, 283
    Extrude button, 275-276
    Group button, 283
    Move button, 270-272
    Primitives mini-toolbar, 270
    Scale button, 273-274
    Sketch button, 275-276
    Transform button, 275
    Ungroup All button, 283
    Ungroup button, 283
  navigation bar, 276-277
    Fit button, 277
    Orbit button, 278-280
    Pan button, 278
    View Cube button, 281-282
    Zoom button, 278-279
  user interface, 267-269
.123dx file type extension, 269

A
About menu, 36
accounts
  creating, 32
  Thingiverse accounts, 200
Actions menu
  Delete option, 40
  Duplicate option, 40
  Properties option, 40-42
Add Photo button, 38
Adjust button, 76
adjusting workspace dimensions, 255-256
advantages of 3D modeling, 13-14
advantages of Tinkercad, 2
Align feature, 73-83
aligning objects, 73-83, 147-151
alternatives to Tinkercad
  123D Design, 243-244
  FreeCAD, 245-246
  SketchUp, 244-245
animated movies, 3D modeling in, 11, 14
applications of 3D modeling, 14-15
  animated movies, 11, 14
  architecture, 13
  product design, 13
  video games, 11-14
applications. *See specific applications*
architectural applications of 3D modeling, 13
assembling
  Pinewood Derby model body shapes, 257-259
  rocket model, 86-98
    additional tweaks, 98-99
    grouping fins, 96-98
    launchpad scaffolding, 70-85
    raising main body, 87-89
    rotating fins, 89-96

Autodesk
Autodesk 360, 2
123D Catch, 226-227
  editing 3D models, 235-238
  saving STL file, 239-241
  taking photos, 227-234
123D Design, 243-244, 267
  main toolbar, 269-276
  navigation bar, 276-282
  user interface, 267-269
purchase of Tinkercad, 2
axes
explained, 17-21
origin, 19-21
rotating objects around, 89-96
X axis, 18
Y axis, 19

B

Basic Dog Tag Oval - No Text model. *See* dog tag model
benefits of 3D modeling, 13-14
benefits of Tinkercad, 2
Biehler, John, 283
Bishop Mold Container
  aligning objects, 147-151
  copying chess piece model to Dashboard, 132-137
  creating mold halves, 143-146
  creating Play-Doh mold, 138-143
  grouping objects, 152-154
Blender, 11
bonus project. *See* Pinewood Derby racer model
brainstorming, 102-104

C

CAD (computer-aided design), 10
car model. *See* Pinewood Derby racer model
Cartesian coordinate system, 19
"carving" Pinewood Derby model block, 259
  engine, 266
  final shape, 264-265
  Side hole object, 262-264
  side mufflers, 266
  Top hole object, 259-261
  wheels, 265
categories (Gallery)
  Hot Now, 130
  Newest Things, 130
  Staff Favorites, 130
  #template, 131

changing
  color, 52
  properties, 40-42
  units of measurement, 104-105
chess piece model, copying to Dashboard, 132-137
Collections list, 36
color, changing, 52
Combine button (123D Design), 283
computer-aided design (CAD), 10
Continue Printing button, 213
controls, 47-49
  measurement controls, 49-51
  Mirror controls, 177-181
    creating symmetrical pairs of objects, 186
    reversing objects, 181-185
    undoing actions, 183
  Raise/Lower, 72, 87-89
  rotate controls, 47-49
  Rotation, 89-96
Convert File button, 253
converting
  JPEG images to SVG files, 253
  objects to hole objects, 57-66
  real objects to digital models, 225-227
    editing 3D models, 235-238
    saving STL file, 239-241
    taking photos, 227-234
coordinates
  Cartesian coordinate system, 19
  polar coordinates, 19
copying
  models to Dashboard, 133-137
  objects, 66
Copy option (Edit menu), 66
Copy & Tinker button (Gallery), 133-137
costs
  of 3D modeling, 13-14
  of 3D printers, 156
  Tinkercad pricing plans, 32
Create New Design button, 36
cubehero.com, 209
curved edges, creating for dog tag model, 108-112
cylinder object, adding to dog tag model, 108-112

D

Dashboard, 33-35
  copying models to, 133-137
  Gallery button, 130
Delete button (123D Catch), 236
Delete option (Actions menu), 40
deleting models, 40

Design menu, Order a 3D Print option, 212
Digitizer (MakerBot), 225
dog tag model
  basic tag shape, creating, 104-118
    curved edges, 108-112
    grouping objects, 112-113
    hole object, 112-115
    initial rectangle, 106-108
    Snap Grid option, 105-106
    units of measurement, 104-105
  brainstorming ideas, 102-104
  hole objects, 126
  naming, 115-117
  number objects, 127
  raised edge, 119-122
  raised text, 122-126
  reversed edge, 126
  suggested improvements, 126-127
  textured edge, 126
Download All Files button (Thingiverse), 204
Download for 3D Printing dialog, 161
downloading models from Thingiverse, 204
Download This Thing button (Thingiverse), 204
dragging objects onto workspace, 47
Duplicate option (Actions menu), 40
duplicating models, 40

E

edges (dog tag model)
  curved edges, 108-112
  raised edge, 119-122
  reversed edge, 126
  textured edge, 126
Edit/Download button (123D Catch), 233
Edit Grid button, 104
editing 3D models, 235-238
Edit menu
  Copy option, 66
  Grid Properties option, 255
  Paste option, 66
Edit Profile button, 36
enabling WebGL, 31
engine, adding to Pinewood Derby racer model, 266
Extrude button (123D Design), 275-276
extruding objects in 123D Design, 275-276

F

fabster.com, 209
Fane, Bill, 283
filament (plastic), 165-167
files, 188-190
  .123dx file type extension, 269
  JPEG files, converting to SVG, 188-190, 253
  STL files
    creating, 161-165
    importing, 187-194
    saving in 123D Catch, 239-241
  SVG files
    converting to JPG, 188-190
    importing, 187-194
finding
  models
    additional 3D model sources, 209
    Thingiverse, 199- 207
  Tinkercad, 30-34
fine-tuning 3D printers, 174
fins (rocket model)
  creating, 57-66
  grouping, 96-98
  rotating, 89-96
Fit button (123D Design), 277
forums, FreeCAD, 245
FreeCAD, 245-246
freecadweb.org, 245

G

G-code, 175
Gallery
  copying models to Dashboard, 133-137
  Gallery menu, 35
  Hot Now category, 130
  previewing models in, 132-133
  Staff Favorites category, 130
  #template category, 131
  viewing models in, 130
Google, affiliation with SketchUp, 244
grabcad.com/library, 209
grid
  properties, 104, 255
  Snap Grid option, 105-106, 256
  units of measurement, 104-105
Grid Properties dialog, 104
Grid Properties option (Edit menu), 255
Group button, 63, 261, 264, 283
grouping objects, 61-64, 96-98, 112-113, 152-154, 261

H

Hole button, 59, 263
hole objects
  adding to dog tag model, 112-115, 126
  adding to Pinewood Derby racer model
    Side hole object, 262-264
    Top hole object, 259-261
  converting objects to, 57-66
Hot Now category (Gallery), 130

I

ID tag model. *See* **dog tag model**
images, converting to SVG files, 253
i.materialize, 212-217
importing
  objects into *Minecraft*, 217-223
  sketches, 187-194
  STL files, 187-194
  SVG files, 187-194
  Thingiverse models, 205
interface (123D Design), 267-269

J-K-L

JavaScript, 195-197
JPEG images, converting to SVG files, 188-190, 253

launchpad (rocket model)
  launchpad objects, 45-53
  scaffolding assembly, 70-85
    aligning objects, 73-79, 83
    stacking objects, 70-72
Learn menu, 35
lessons (Tinkercad), 43-44
licenses (models), 202-204
License menu, 42
Like button, 38
limitations of Tinkercad, 243

M

main screen (123D Design), 267-269
main toolbar (123D Design), 269-270
  Combine button, 283
  Extrude button, 275-276
  Group button, 283
  Move button, 272
  Primitives mini-toolbar, 270
  Scale button, 273-274
  Sketch button, 275-276
  Transform button, 275
  Ungroup All button, 283
  Ungroup button, 283
MakerBot, 199
MakerBot Digitizer, 225
Maker Faire, 2
MCEdit, 220-223
measurement units, changing, 104-105
measurement controls, 49-51
***Minecraft*, 11, 14**
  importing objects into, 217-223
Mirror controls, 177-181
  creating symmetrical pairs of objects, 186
  reversing objects, 181-185
  undoing actions, 183
mirroring objects, 177-186
model repositories
  additional 3D model sources, 209
  Thingiverse, 199-207
    creating accounts, 200
    downloading models, 204
    importing models into Tinkercad, 205
    model license limitations, 202-204
    searching for models, 200
    uploading models to, 206-207
    viewing model details, 200-204
models. *See also* **objects**
  Bishop Mold Container
    aligning objects, 147-151
    copying chess piece model to Dashboard, 132-137
    creating mold halves, 143-146
    creating Play-Doh mold, 138-143
    grouping objects, 152-154
  brainstorming ideas, 102-104
  converting real objects to, 225-227
    editing 3D models, 235-238
    saving STL file, 239-241
    taking photos, 227-234
  copying to Dashboard, 133-137
  definition of, 8
  deleting, 40
  dog tag
    basic tag shape, creating, 104-118
    brainstorming ideas, 102-104
    hole objects, 126
    naming, 115-117
    number objects, 127
    raised edge, 119-122
    raised text, 122-126
    reversed edge, 126
    suggested improvements, 126-127
    textured edge, 126

downloading from Thingiverse, 204
duplicating, 40
editing in 123D Catch, 235-238
finding
   additional 3D model sources, 209
   Thingiverse, 199-207
Gallery
   copying models to Dashboard, 133-137
   Hot Now category, 130
   Newest Things, 130
   previewing models in, 132-133
   Staff Favorites category, 130
   #template category, 131
   viewing models in, 130
importing
   into Minecraft, 217-223
   from Thingiverse, 200, 205
license limitations, 202-204
molds, creating. *See* Bishop Mold Container
naming, 115
Pinewood Derby racer
   background, 247, 249
   "carving" the block, 259-266
   creating body shapes, 250-259
Preview window, 37-38
printing, 155-156
   3D printer costs, 156
   3D printer overview, 156-160
   3D printing services, 212-217
   3D printing software, 171-175
   fine-tuning, 174
   nozzle movement, 167-171
   plastic filament, 165-167
   printer motors, 167-171
   STL files, creating, 161-165
   summary, 175-176
properties, changing, 40-42
renaming, 40-42, 135
rocket model
   launchpad, 45-53
   launchpad scaffolding assembly, 70-85
   rocket assembly, 86-99
   rocket body, 53-56
   rocket fins, 57-66
saving, 56
starting, 36
uploading to Thingiverse, 206-207
modifying shapes, 54-55, 59
molds, creating. *See* Bishop Mold Container
More Info on Creative Commons Licenses link, 42
motors (3D printer), 167-171
Move button (123D Design), 272
moving
   3D printer nozzle, 167-171
   objects, 51, 272
multiple parts, selecting, 61-64

N

naming
   dog tag model, 115-117
   models, 115
   objects, 117
navigating
   123D Design, 276-277
      Fit button, 277
      Orbit button, 278-280
      Pan button, 278
      View Cube button, 281-282
      Zoom button, 278-279
   Tinkercad, 35-39
navigation bar (123D Design), 276-277
   Fit button, 277
   Orbit button, 278-280
   Pan button, 278
   View Cube button, 281-282
   Zoom button, 278-279
Newest Things category (Gallery), 130
nozzle movement (3D printers), 167-171
number objects (dog tag model), 127

O

objects. *See also* **models**
   aligning, 73-83, 147-151
   color, changing, 52
   controls, 47-49
      measurement controls, 49-51
      rotate controls, 47-49
   converting real objects to digital models, 225-227
      editing 3D models, 235-238
      saving STL file, 239-241
      taking photos, 227-234
   converting to hole objects, 57-66
   copying, 66
   cylinder, adding to dog tag model, 108-112
   dragging onto workspace, 47
   extruding in 123D Design, 275-276
   grouping, 61-64, 96-98, 112-113, 152-154, 261
   hole objects
      adding to Pinewood Derby racer model, 259-264
      converting shapes to, 57-66
      dog tag model, 112-115
   importing into *Minecraft*, 217-223
   mirroring, 177-186
   modifying shape of, 54-55, 59
   moving, 51, 272
   multiple parts, selecting, 61-64
   naming, 117
   pasting, 66

objects

raising, 87-89
rectangle, adding to dog tag model, 106-108
resizing, 49-51, 271-275
reversing, 181-185
rotating, 25-28, 89-96, 272
stacking, 70-72
symmetrical pairs of objects, creating, 186
one-dimensional (1D) points, 8
online-convert.com, 188, 253
opening
　new workspaces, 36
　Preview window, 37-38
　Tinkercad, 30-34
Orbit button (123D Design), 278-280
Orbit tool (123D Catch), 235
Order a 3D Print option (Design menu), 212
origin, 19-21
orthogonal view, 9

P

Pan button (123D Design), 278
Paste option (Edit menu), 66
pasting objects, 66
photos
　adding to Tinkercad accounts, 34
　taking in 123D Catch, 227-234
Pinewood Derby racer model
　background, 247-249
　"carving" the block
　　engine, 266
　　final shape, 264-265
　　Side hole object, 262-264
　　side mufflers, 266
　　Top hole object, 259-261
　　wheels, 265
　creating body shapes
　　assembling shapes, 257-259
　　images, converting to SVG files, 253
　　objects, resizing, 256
　　Side template, 251-253
　　Top template, 250-251
　　workspace dimensions, adjusting, 255-256
Plane Cut tool (123D Catch), 237
planes, 22-25
plastic filament, 165-167
Play-Doh mold
　aligning objects, 147-151
　creating, 138-143
　grouping objects, 152-154
　mold halves, 143-146
polar coordinates, 19
Ponoko, 212
previewing models in Gallery, 132-133
Preview window, 37-38, 133

pricing plans (Tinkercad), 32
Primitives mini-toolbar (123D Design), 270
printers (3D)
　controlling with printing software, 171-175
　costs, 156
　explained, 156-160
　fine-tuning, 174
　MakerBot Replicator 2, 159-160
　motors, 167-171
　nozzle movement, 167-171
　plastic filament, 165-167
　Printrbot Simple, 158
　RepRap Darwin, 158-159
　summary, 175-176
printing models, 155-156
　3D printers
　　controlling with printing software, 171-175
　　costs, 156
　　explained, 156-160
　　fine-tuning, 174
　　MakerBot Replicator 2, 159-160
　　motors, 167-171
　　nozzle movement, 167-171
　　plastic filament, 165-167
　　Printrbot Simple, 158
　　RepRap Darwin, 158-159
　3D printing services, 212-217
　printing software, 171-175
　STL files, creating, 161-165
　summary, 175-176
printing software, 171-175
Printrbot Simple, 158
Private setting, 39
product design, 3D modeling in, 13
programmers, 194
Projects list, 36
projects. *See* **models**
properties, changing, 40-42
Properties option (Actions menu), 40-42

R

raised edge (dog tag model), creating, 119-122
raised text (dog tag model), creating, 122-126
Raise/Lower control, 72, 87-89, 178
raising objects, 87-89
real objects, converting to digital models, 225-227
　editing 3D models, 235-238
　saving STL file, 239-241
　taking photos, 227-234
rectangle object
　adding to dog tag model, 106-108
　sizing for dog tag model, 106
Redo button, 96
renaming models, 40-42, 135

repables.com, 209
Repetier, 172
Replicator 2, 159-160
repositories, Gallery, 130
  copying models to Dashboard, 133-137
  Hot Now category, 130
  previewing models in, 132-133
  Staff Favorites category, 130
  #template category, 131
  viewing models in, 130
RepRap Darwin, 158-159
resizing objects, 49-51, 256, 271-275
reversed edge (dog tag model), creating, 126
reversing objects, 181-183, 185
rocket model
  launchpad
    launchpad objects, 45-53
    scaffolding assembly, 70-85
  rocket assembly, 86-98
    additional tweaks, 98-99
    grouping fins, 96-98
    raising main body, 87-89
    rotating fins, 89-96
  rocket body, 53-56
  rocket fins, 57-66
rotate controls, 47-49
rotation
  explained, 25-28
  objects, 89-96, 272
  workspaces, 47-49
Rotation controls, 89-96

S

sample lessons (Tinkercad), 43-44
Save button, 56
Save Changes button, 42, 117
saving
  STL files in 123D Catch, 239-241
  work, 56
scaffolding (rocket model), assembling, 70-85
  aligning objects, 73-79, 83
  stacking objects, 70-72
scalable vector graphics. *See* SVG files
Scale button (123D Design), 273-274
Sculpteo, 212
Search box, 35
searching Thingiverse, 200
Select button (123D Catch), 235
selecting multiple parts, 61-64
Select Photos button (123D Catch), 231
services, 3D printing services, 212-217
Shape Generators tool, 194-196

shapes
  basic dog tag shape, creating, 104-118
    curved edges, 108-112
    grouping objects, 112-113
    hole object, 112-115
    initial rectangle, 106-108
    Snap Grid option, 105-106
    units of measurement, 104-105
  modifying, 54-55, 59
  Pinewood Derby model body shapes
    assembling, 257-259
    images, converting to SVG files, 253
    objects, resizing, 256
    Side template, 251-253
    Top template, 250-251
    workspace dimensions, adjusting, 255-256
Shapeways, 212
Show More button (Gallery), 132
Side hole object (Pinewood Derby racer model), 262-264
side mufflers, adding to Pinewood Derby racer model, 266
Side template (Pinewood Derby model), 251-253
Sign In/Join button (Thingiverse), 200
Sign Up for Free Account button, 32
Sketch button (123D Design), 275-276
sketches, importing, 187-194
SketchUp, 244-245
sketchup.com, 244
Snap Grid, 105-106, 256
stacking objects, 70-72
Staff Favorites category (Gallery), 130
Standard view button, 38
Start a New Project button (123D Catch), 228
STL files
  creating, 161-165
  importing, 187-194
  saving in 123D Catch, 239-241
SVG files
  converting images to, 253
  converting to JPG, 188-190
  importing, 187-194
symmetrical pairs of objects, creating, 186

T

taking photos in 123D Catch, 227-234
Team Fortress 2, 12
#template category (Gallery), 131
templates (Pinewood Derby model)
  Side template, 251-253
  Top template, 250-251
text (dog tag model), 122-126
textured edge (dog tag model), 126

Thingiverse, 199-207
  creating accounts, 200
  downloading models, 204
  importing models into Tinkercad, 205
  model license limitations, 202-204
  searching for models, 200
  uploading models to, 206-207
  viewing model details, 200-204
Thing Properties dialog, 115
Tinker This button, 38-40
toolbar (123D Design), 269-270
  Combine button, 283
  Extrude button, 275-276
  Group button, 283
  Move button, 272
  Primitives mini-toolbar, 270
  Scale button, 273-274
  Sketch button, 275-276
  Transform button, 275
  Ungroup All button, 283
  Ungroup button, 283
Top hole object (Pinewood Derby racer model), 259-261
Top template (Pinewood Derby model), 250-251
Toy Story, 11
Transform button (123D Design), 275
Trimble SketchUp, 244-245
two-dimensional (2D) objects, 8

U

Undo button, 63, 96, 183
undoing actions, 96
Ungroup All button (123D Design), 283
Ungroup button (123D Design), 283
units of measurement, changing, 104-105
Update Grid button, 104
Upload a Thing! button (Thingiverse), 206
uploading models to Thingiverse, 206-207
user accounts, creating, 32

V

video games, 3D modeling in, 11-14
View 3D button, 38
View Cube button (123D Design), 281-282
viewing
  models in Gallery, 130
  Thingiverse model details, 200-204

W

WebGL, enabling, 31
websites, 245
  3dhacker.com, 209
  123dapp.com/design, 244
  cubehero.com, 209
  fabster.com/physibles, 209
  grabcad.com/library, 209
  makerfaire.com, 2
  online-convert.com, 188, 253
  repables.com, 209
  sketchup.com, 244
  Thingiverse, 199-207
    creating accounts, 200
    downloading models, 204
    importing models into Tinkercad, 205
    model license limitations, 202-204
    searching for models, 200
    uploading models to, 206-207
    viewing model details, 200-204
  youmagine.com, 209
wheels, adding to Pinewood Derby racer model, 265
workspaces
  adjusting dimensions of, 255-256
  dragging objects onto, 47
  opening new, 36
  rotating, 47-49
  zooming in/out, 45-47

X-Y-Z

X axis, 18
X motors (3D printers), 168

Y axis, 19
Y motors (3D printers), 168-169
youmagine.com, 209

Z motors (3D printers), 169,-170
Zoom button (123D Design), 278-279
zooming in/out, 45-47, 70
Zoom Out feature, 70

Other Books
YOU MIGHT LIKE!

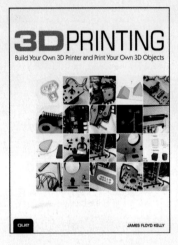

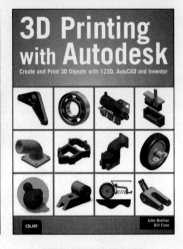

ISBN: 9780789752352   ISBN: 9780789748836   ISBN: 9780789753281

SAVE 30%
Use discount code **HARDWARE**

Visit **quepublishing.com** to learn more!

* Discount code HARDWARE is valid for a 30% discount off the list price of eligible titles purchased on informit.com or quepublishing.com. Coupon not valid on book + eBook bundles. Discount code may not be combined with any other offer and is not redeemable for cash. Offer subject to change.

ALWAYS LEARNING   PEARSON

QUEPUBLISHING.COM
Your Publisher for Home & Office Computing

Quepublishing.com includes all your favorite—and some new—Que series and authors to help you learn about computers and technology for the home, office, and business.

Looking for tips and tricks, video tutorials, articles and interviews, podcasts, and resources to make your life easier? Visit **quepublishing.com**.

- **Read the latest articles and sample chapters** by Que's expert authors

- **Free podcasts** provide information on the hottest tech topics

- **Register your Que products** and receive updates, supplemental content, and a coupon to be used on your next purchase

- **Check out promotions and special offers** available from Que and our retail partners

- **Join the site** and receive members-only offers and benefits

QUE NEWSLETTER
quepublishing.com/newsletter

 twitter.com/quepublishing

 facebook.com/quepublishing

 youtube.com/quepublishing

 quepublishing.com/rss

 Que Publishing is a publishing imprint of Pearson

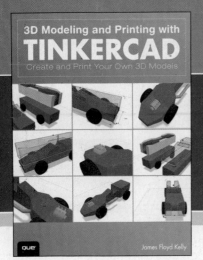

Safari Books Online

FREE Online Edition

Your purchase of *3D Modeling and Printing with Tinkercad* includes access to a free online edition for 45 days through the **Safari Books Online** subscription service. Nearly every Que book is available online through **Safari Books Online**, along with thousands of books and videos from publishers such as Addison-Wesley Professional, Cisco Press, Exam Cram, IBM Press, O'Reilly Media, Prentice Hall, Sams, and VMware Press.

Safari Books Online is a digital library providing searchable, on-demand access to thousands of technology, digital media, and professional development books and videos from leading publishers. With one monthly or yearly subscription price, you get unlimited access to learning tools and information on topics including mobile app and software development, tips and tricks on using your favorite gadgets, networking, project management, graphic design, and much more.

Activate your FREE Online Edition at
informit.com/safarifree

STEP 1: Enter the coupon code: SRCUPVH.

STEP 2: New Safari users, complete the brief registration form. Safari subscribers, just log in.

If you have difficulty registering on Safari or accessing the online edition, please e-mail customer-service@safaribooksonline.com